M000318173

Collective Animal Behavior

Collective Animal Behavior

DAVID J. T. SUMPTER

PRINCETON UNIVERSITY PRESS

Princeton and Oxford

Copyright © 2010 by Princeton University Press

Published by Princeton University Press, 41 William Street,

Princeton, New Jersey 08540
In the United Kingdom: Princeton University Press, 6 Oxford Street,
Woodstock, Oxfordshire OX20 1TW

All Rights Reserved

Library of Congress Cataloging-in-Publication Data

Sumpter, David J. T., 1973–
Collective animal behavior / David J. T. Sumpter.
p.   cm.
Includes bibliographical references and index.
ISBN 978-0-691-12963-1 (hardcover : alk. paper)
ISBN 978-0-691-14843-4 (pbk : alk. paper)
1. Social behavior in animals. 2. Collective behavior. I. Title. QL775.S86 2010
591.56—dc22
2009053055

British Library Cataloging-in-Publication Data is available

This book has been composed in Sabon

Printed on acid-free paper. ∞

press.princeton.edu

Printed in the United States of America

1 3 5 7 9 10 8 6 4 2

To Lovisa, Henry, and Elise

Contents

Acknowledgments

It is collaboration that allows me to learn and inspires me to work. I enjoy the collective effort involved in the type of research I do. It is probably the importance of social interactions in my own life that led me to the research topic of this book. Is it then a contradiction to this view of the world that I chose to shut myself up and write a book by myself? I don't think so. Although I like working with others, anyone who works with me can confirm that I can be slightly single minded about how I do things. Writing this book has provided an outlet for my own impressions and ideas, which has hopefully allowed me to keep a better balance in my collaborations with others.

There are five biologists, all of whom I met during my final year as a doctoral student, who have greatly contributed to the thinking on which this book is based. Stephen Pratt, Madeleine Beekman, Iain Couzin, Max Reuter, and Dora Biro have in very different ways taught me how to think as a biologist, as well as provided me with endless hours of conversation, entertainment, gossip, intrigue, and downright drunken scandal that has enriched my working life over the last 10 years.

There are two people who have shaped the way I think about mathematics and modeling and I whom would like to thank especially. Having Dave Broomhead as a PhD supervisor taught me a way of approaching the world that has stayed with me ever since and, I hope, is reflected in every mathematical box in this book. Many conversations with Ander Johansson have contributed not only to my understanding of mathematics, but also to its possibilities and limitations in describing science.

Many other people have either commented directly on the text of this book or have, possibly unwittingly, contributed to it through conversation, email exchange, or the odd sentence stolen from a joint paper. These include: Åke Brännström, Jerome Buhl, Lisa Collins, Larissa Conradt, Jean-Louis Deneubourg, Audrey Dussutour, Kevin Foster, Nigel Franks, Deborah Gordon, Michael Griesser, Joe Hale, Peter Hedström, Kerri Hicks, David Hughes, Duncan Jackson, Neil Johnson, Christian Jost, Alex Kacelnik, Jens Krause, Laurent Lehmann, Esther Miller, Stam Nicolis, Andrea Perna, Sophie Persey, Francis Ratnieks, Steve Simpson, Guy Theraulaz, Ashley Ward, Jamie Wood and Kit Yates. A special thank you to Graeme Ruxton and the two other anonymous reviewers for their highly constructive comments on various drafts of the book.

ACKNOWLEDGMENTS

Thank you to Qi Ma, Etsuko Nonaka, and Daniel Strömbom for their detailed reading of the book and participating in an intensive course based on its contents. A particular thank you to Boris "you can't inflict this on the world" Granovskiy for both participating in the course and then proofreading the entire book (apart from the acknolwedgements). Thank you to Joachim Munkhammar for creating the webpages for the book and, together with Qi Ma, Kit Yates, and Stam Nicolis, for contributing to some of the simulations in the modeling boxes.

I wrote most of this book during the winter of 2006 and spring of 2007. During this time I worked half-time and the other half of my time I was on parental leave. Thank you, Elise and Henry, for always forgiving my preoccupation with other matters when I should have been fully dedicated to you. Thank you to Lovisa, for her love, support, patience, and inspiration. Thanks also to my mum, dad, and family for their constant support.

Thank you to Sam Elworthy, Alison Kalett, Janie Chan, Stefani Wexler, Sara Lerner, Lor Gehret, and all the other staff at Princeton University Press for all their work in preparing this book for publication.

Lastly, I would like to thank the Royal Society, whose flexible and long-term funding gave me the time and confidence to write the first half of this book, and the mathematics department at Uppsala University, for providing me with such a wonderful working environment while I was getting through the second half.

—Chapter 1—

Introduction

Some scientists invest an entire career in the study of organisms of a single species, others in understanding particular types of cells or in determining the role of a certain gene. The elements of each level of biological organization can take more than a lifetime to understand. How then can we put all this information together? Understand how genes interact to drive the cell, how cells interact to form organisms, and how organisms interact to form groups and societies? These questions are fundamental to the scientific endeavor: how do we use our understanding of one level of organization to inform us about the level above?

Linking different levels of organization involves the study of collective phenomena: phenomena in which repeated interactions among many individuals produce patterns on a scale larger than themselves. Collective phenomena are within us and all around us: the clustering of cells to build our bodies, the firing of neurons in our brains, flocks of birds twisting above our heads, and the pulsating mass of bodies surrounding us on a Saturday night dance floor. Understanding these phenomena is an important part of the fields of developmental biology, neuroscience, behavioral ecology, and sociology, to name just a few. Even researchers studying the most intricate details of the components of a particular system are acutely aware of the need to understand how these components fit together to create a whole system.

The study of collective phenomena is founded on the idea that a set of techniques can be applied to understand systems at many different physical scales. This idea originated from mathematics, theoretical physics, and chemistry. Books by Wiener (1948), Ashby (1947), von Bertalanffy (1968) and Nicolis & Prigogine (1977) all aimed at providing a framework for the study of collective phenomena. Von Bertalanffy argued for the existence of general growth laws of social entities as diverse as manufacturing companies, urbanization, and Napoleon's empire. Wiener argued that homeostasis, a stable functioning of natural systems, could be achieved through simple feedback loops. Nicolis and Prigogine aimed to pin down a rigorous theory of non-linear thermodynamics, explaining similarities between systems at very different scales. For example, could the flow of traffic be described by the mathematics of fluid flow? And if this were the case could we make general statements about the flow of any type of matter, be it swarms of locusts, crowds leaving football grounds, or water running down the drain?

Over the last 30 years, research into collective phenomena and complex systems has rapidly expanded. The completion of the human and other genome projects was followed by a call for systems biology, a combination of experimental and computational approaches to integrate our collected database of biological facts (Kitano 2002). The challenges of assessing changes in the global environment require understanding of ecological interactions that occur over many different temporal and spatial scales (Levin 1992, 2000; Stainforth et al. 2005). Study of neurobiology and the immune system again involve understanding how neurons or cells interact to make decisions (Bays & Wolpert 2007; George et al. 2005). Modern sociology and social psychology aim to link the decisions of individuals to the social behavior of the many (Hedstrom 2005; Milgram 1992; Schelling 1978). Accompanying the realization of the importance of collective phenomena within different fields has been the development of mathematical modeling tools for investigating these systems.

Animal groups provide many key examples of collective phenomena. They also provide some of the most spectacular and fascinating sights in the natural world (figure 1.1). Flocks of birds turning in unison or migrating in well ordered formation; fish shoals splitting and reforming as they outmaneuver a predator; swarms of honeybees settled on a branch of a tree while colony members use dances to debate where they will fly; and the long bifurcating trails along which ants transport food and materials have all long fascinated scientists. There is something captivating about the patterns these groups create. They are neither entirely regular, nor are they entirely random. They are, quite simply, complex.

This book is about such collective animal behavior. It is a study of how interactions between animals produce group level patterns and why these interactions have evolved. It is aimed at two different types of reader: at the behavioral ecologist who is interested in how techniques for studying collective phenomena and complex systems are applied to animal groups; and at the general scientist who would like to see solid examples of the application of techniques for the study of collective phenomena. I will start by arguing, for the sake of the bookshop browser who has at least read this far, why you should now buy this book.

Why Collective Animal Behavior?

The study of collective animal behavior gives a key example of how experiment can be combined with a theory of complex systems to better understand the world around us. As we have gained empirical knowledge of group behavior across species there has been an increasing need for theoretical concepts, as well as mathematical and simulation models, that allow us to unify this knowledge into a general understanding. This book describes such concepts and tools, and shows how they can be applied.

Far from being a book solely about mathematical models, an equal focus is made on empirical examples. Each chapter describes a large number of different biological systems, and how one or two models can be applied in aiding our understanding of these different systems. This approach is taken in order to demonstrate the applicability of the theory to experiment and observation. Furthermore, by showing how similar models can be applied to very different systems I aim to demonstrate the logical relationships that can be drawn between very distinct biological systems. Modeling is a tool for understanding general properties of different systems.

Animal behavior provides a wealth of interesting and accessible examples of collective phenomena and complex systems. Most people are familiar with ant trails; cockroach aggregations; fish schools; bird migrations; honeybee swarms; web construction by spiders; and locust marching, even if they have not observed them personally. In these systems there are two clearly defined levels of organization that we aim to link together: the animal and the group. This clarity stands in contrast to many other collective phenomena, such as protein interactions or ecological webs, where it can be difficult to establish exactly on which level to observe a system. Thus animal behavior provides much needed case studies of how complex systems theory can be put into practice.

What the study of animal behavior might gain in terms of accessibility, it loses in terms of experimental precision. Animals are intrinsically more complicated than proteins or cells and it can be difficult to provide a clean description of the behavior of individuals. Individual variation and difficulty in collecting large numbers of replicates means that we can seldom write down all-encompassing mathematical models for the actions and interactions of animals. The study of collective animal behavior tests to the limit the supposed unifying nature of mathematical modeling. Individual animals are inherently difficult to predict, but can we still make strong predictions about their collective behavior? The answer is that in many situations we can. Applied correctly, mathematical models are able

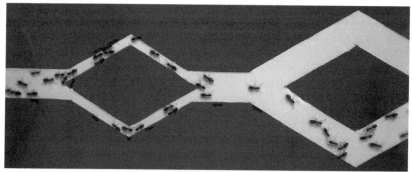

(a)

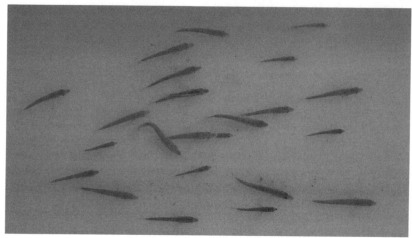

(b)

(c)

(d)

(e)

(f)

Figure 1.1 The collective life of animals. (a) The flow of traffic of *Lasius niger* ants between nest and food have been studied both in terms of their recruitment (chapter 3) and congestion (chapter 8). Copyright: Daniel Perrin, CNRS, phototeque. (b) The decision-making of sticklebacks reveals that they can reach consensus about their direction of travel (chapter 4). Copyright: Jolyon Faria. (c) Locusts form large coherent marching bands despite only local interactions (chapter 5). Copyright: Iain Couzin. (d) Homing pigeons compromise in finding their route home, but only if conflict between their directional information is low (chapter 5). Copyright: Dora Biro. (e) Honeybees, in this case *Apis florea,* form bivouac structures (chapter 6) while they decide where they should move for a new home (chapter 10). Copyright: James Makinson. (f) Temnothorax ants communicate during their search for a new home by leading tandem runs between prospective nest sites (chapter 10). Copyright: Stephen Pratt.

to make predictions about the group behavior of even the most complex of animals, including humans.

The study of collective animal behavior also allows us to better understand the various approaches to studying biology. The two main approaches are known as mechanistic and functional. The mechanistic approach looks at *how* animals interact to produce group level patterns. This approach concentrates on identifying communication mechanisms, such as visual and chemical signals, and tries to determine how these mechanisms are integrated to produce collective patterns. Traditionally, the study of collective phenomena and complex systems is more closely associated with such mechanistic explanations.

Functional explanations are based on arguments about *why* a behavior evolved through natural selection: animals with behaviors that improve their chances of reproduction will increase in frequency in the population, and those with behaviors detrimental to survival will die out. Thus natural selection acts to produce behaviors consistent with the selfish interests of the individual or, more properly, the genes carried by the individual (Dawkins 1976, 1982). Functional explanations take a central role in understanding why individuals co-operate to form collective patterns, why these patterns persist despite the conflicting interests of the individuals creating them, and why some collective patterns are inconsistent with the selfish individual or selfish gene.

Collective animal behavior provides an excellent opportunity to study the link between function and mechanism. While these two forms of explanation are complementary, they are also interdependent. We cannot understand why co-operation evolves without knowing the mechanisms by which co-operative patterns are generated. Nor can we study mechanisms without considering the potential conflicts that can arise between individuals. This book seeks to clarify the relationship between mechanistic and functional explanations, by providing both sorts of explanation and showing how they fit together.

Mathematical Modeling

The study of collective phenomena goes hand in hand with the use of mathematical and simulation modeling. The arguments made in this book often rely on mathematics, and include applications of a wide range of mathematical models. What all of these models have in common is that they are logical arguments, taking us from one set of statements about the real world to another. Mathematics is a way of traveling logically from A to B, where A is a set of precisely stated rules or assumptions that are thought to characterize a system and B is a set of predictions logically arising from these properties. Mathematics is a language for making precise statements about the world and following their consequences to a logical conclusion (Feynman 1965).

In its idealized form, the mathematical modeling cycle proceeds as follows: (1) by formalizing some of our knowledge about a system in a mathematical model we generate a set of assumptions A, (2) we use mathematical analysis or computer simulations to make a set of predictions B, (3) we confirm or refute these predictions against our available knowledge and against the outcome of new experiments, and (4) we return to step 1 and revise our assumptions in light of the outcome of the model and any new experiments. In theory, the cycle converges to a more and more accurate picture of reality.

Rather then being a strictly defined procedure, however, the modeling cycle is a convenient way of summarizing a whole range of mathematical and scientific activities. We are seldom in a position to consider or refine only a single set of assumptions or predictions. Instead, many different sets of assumptions can all have consequences consistent with the known relevant properties of a system. In such situations, modeling can suggest experiments that can discern which of several alternatives is the most accurate description of a system. It is often when we propose a model, make predictions, and find that these predictions are wrong that modeling provides its greatest insights into the real world.

Increased understanding of natural systems also arises from simply playing with mathematical models. By looking at how different assumptions lead to different predictions and comparing these outcomes with what we know about a system, we can ensure that our own understanding of a system is logically consistent. Thus a great deal of worthwhile modeling remains at the stage of dividing our understanding of a system into the empirical observations we take to be assumptions and the observations we label as predictions. For example, we might ask the question: what is the minimal set of assumptions we can make that predicts all known system properties? This application of Occam's razor serves to clarify and condense our understanding of a system in terms of its basic underlying principles.

It must be borne in mind that any particular mathematical model does not provide a unique way of thinking about the world. It is often the case that two completely different mathematical descriptions of a system are entirely compatible with each other. In particular, we can make two different sets of assumptions about a system, follow both of them to their logical conclusion, and produce two entirely different predictions. Provided the assumptions do not contradict one another, then neither do

7

the predictions necessarily contradict each other. Remembering this apparently innocuous fact can resolve a lot of arguments about which of a variety of models of a system is "best."

From Individuals to Collectives

Models of collective animal behavior are often based on assumptions about the behavior at the level of individuals and then are used to make predictions about the patterns created at the level of the group. For example, we might describe how individual ants leave and follow chemical trails and predict the collective structure of their trail network. This division between the individual and the group leads us to expressions like "emergence" and "self-organisation" (Bonabeau et al. 1997; Camazine et al. 2001; Deneubourg & Goss 1989; Holland 1998; Kauffman 1993; Nicolis & Prigogine 1977). We make some relatively simple assumptions about individual behavior from which emerge predictions about group behavior. The group level pattern is said to self-organize because it was not encoded directly in the individual-level rules. Mathematical models are a way of extracting otherwise difficult-to-see connections between the interactions of individuals and the patterns created at a group level. The connection to mathematics removes any mystical meaning or ambiguity that might lurk within phrases such as "emergence" and "self-organization." These phrases serve primarily to highlight the fact that simple interactions between individuals can produce sometimes surprising and empirically testable predictions about collective patterns.

There is a sense in which it is useful to talk about the principles arising from mathematical models of collective behavior (Sumpter 2006). This is where the same model provides insights into many different and seemingly unrelated systems. Every chapter of this book provides an example where the same mathematical model has explanatory power across different biological systems. For example, one model creates connections between firefly flashing and human applause (chapter 6) or another model connects ants foraging for food to cockroaches finding a shelter (chapters 3 and 4). This universal application of certain mathematical models is a remarkable observation. We can pick up the assumptions we have used to describe one system and apply them directly to produce predictions about a second system. Adding to this the fact that we can perturb our model and still produce the same predictions, or see changes in the predictions that reflect differences in the systems, would suggest fundamental laws encoded within these models.

These logical connections between systems prove extremely useful. In this book, I proceed on a case by case basis through different examples of collective animal behavior. For each particular system, I classify how individuals interact with each other and make this the basis of mathematical models. When a similar mathematical model has been previously applied to another system, this helps us understand the behavioral algorithms and thus the system.

Functional and Mechanistic Approaches

Making assumptions about how individuals behave and predicting outcome at a group level is a mechanistic approach to the study of behavior. We collect all the information about how individuals behave in response to their environment and to other individuals and incorporate this detail into a mathematical model that predicts the collective patterns generated by the group. In this way, we attempt to determine the mechanisms through which the collective outcome is formed.

The mechanistic approach to biology is often contrasted with the functional approach. In the functional approach we ask what the reproductive value or function is of a particular behavioral strategy. Mathematical models of function are usually based on assumptions about the costs and benefits, in terms of their impact on survival and reproduction, of a particular strategy. If we can identify the costs and the benefits associated with a strategy and compare them to those associated with an alternative strategy, we can predict how behavior evolves through natural selection. We can also measure the extent to which animals are able to change their behavioral strategy in response to changing costs and benefits.

Functional questions are particularly interesting in the context of more than one individual, because what might be a benefit to one individual can be a cost to another. When natural selection acts to increase the frequency of a particular type of behavioral strategy in the population, it simultaneously changes the cost-benefit relationship for others in the population. As such, instead of acting to maximize some static function, natural selection acts to increase or decrease different types of strategy until an equilibrium is reached where no individual can evolve to do better by changing strategy (Dugatkin & Reeve 1998; Maynard Smith 1982). This picture is complicated by interactions with relatives. If a genotype evolves that helps relatives then it can be selected for, not because it directly increases the survival value of the individual carrying the genotype, but because it increases the survival value of other individuals that also carry the genotype (Hamilton 1964).

Krebs & Davies (1993) provide a number of excellent examples of the difference between functional and mechanistic explanations. For example, female lions living in groups tend to come into oestrus at the same time. Krebs & Davies argue that there are functional benefits to the synchronisation of oestrus cycles. Mother lions suckle their young communally, so cubs born to a group of simultaneously lactating lions will have feeding opportunities even when their mother is out hunting. This benefit is further increased by the fact that the lions are relatives. Related lionesses pass on their genes not only through their own offspring, but also by raising the offspring of their sisters or cousins.

A mechanistic explanation looks at the process through which synchronization occurs. Lionesses' oestrus cycles are known to be coupled by the release of pheromones. When living alone, individual lionesses may have cycles with slightly different periods, but these cycles become entrained through the release of pheromones. Pheromone release by a lioness coming into oestrus will speed up or slow down those that are out of phase with her. Eventually, all lionesses will adopt the same cycle. This mechanistic explanation is well supported by mathematical models of synchronization (Strogatz 2003). These models predict that when group members behave periodically, but each with a different phase and possibly a different period, then, provided there is some means to communicate phase, the members can synchronize their cycles. The mathematical models give predictions about how factors such as individual variation and strength of coupling affect the degree of synchronization.

Despite such clear examples, demarcating mechanistic and functional explanations can be a walk through a minefield. Let us for argument's sake consider a hypothetical study in which it was found that highly related females had synchronized oestrus cycles and less related females did not. It would be tempting to conclude from this study that unrelated females do not benefit from synchronization and thus do not synchronize their cycles. This is a functional prediction consistent with kin selected benefits. However, the mathematical results about synchrony state that if there is too much between-individual variation in the period of cycles, as there may be between less related individuals, then synchronization becomes impossible (see chapter 6). The lack of synchrony may be due simply to a mechanistic "failure" in unrelated groups. This alternative explanation could be partially resolved by, for example, measuring pheromone release or experimentally manipulating group composition to consist of unrelated lions with similar oestrus cycles. Even after these tests there can remain problems in making functional interpretations. For example, did synchrony evolve because of kin selected benefits or do lion females remain in kin groups in order to syncronize their oestrus cycles?

Biologists are aware of these kinds of problems and I am confident that there would be some route through my rather simplified discussion of lion oestrus cycles. My point, however, is to re-emphasize that an increased knowledge of the mechanisms at work in any particular system can change our functional explanation of its behavior. Mechanisms should not simply be considered as a way of obtaining parameters for the cost-benefit curves of functional models. Rather we should aim to form functional explanations that fully account for the underlying mechanisms. This point is particularly relevant where interactions produce highly non-linear patterns. Our intuition is not used to dealing with these outcomes and it becomes easy to miss important aspects of mechanisms that completely change our functional predictions. This book is full of examples of how mechanisms and functional explanations must be considered simultaneously if we are to come to a fuller understanding of group behavior.

A more philosophical conflict sometimes arises between proponents of functional and mechanistic explanations. This conflict centers on the question of which type of explanation is more relevant to understanding biology. Factors influencing survival value are often described as "ultimate," while "proximate" is used to describe those governing mechanisms. These labels can give the impression, and probably reflect the feelings of many behavioral ecologists, that functional questions have a greater importance than mechanisms (Krebs & Davies 1997; West et al. 2007). This observation means that theories about self-organization sometimes sit rather uncomfortably alongside the theory of evolution through natural selection. Throughout this book, I emphasize how the same mechanisms arise again and again in many different systems. Mathematical models formalize these logical connections between systems. For some scientists this gives these models and the principles that underlie them an equal, if not greater, importance than natural selection (Hoelzer et al. 2006; Kauffman 1993; Pepper & Hoelzer 2001; Wolfram 2002).

This book avoids such conflict, which usually obscures the real scientific questions. Whether verbal or mathematical, functional and mechanistic models are based on different assumptions and different predictions. More often than not these assumptions and predictions are consistent with each other. In other cases different decisions have been made about what is relevant or irrelevant in the construction of the models. As emphasized in the mathematical modeling section of this chapter, it is perfectly consistent to have different models of a particular system, each of which makes a different link between assumptions and predictions. My main criterion for the choice between functional or mechanistic explanations in a particular instance is the strength of their explanatory and predictive power. Ideally, both types of explanation should be possible and we should be able to see the link between them.

11

Human Society

One reason why animal groups are such a popular subject for scientific study is the importance of social interactions in our own everyday experience. Humans are inherently social animals, whose activities exhibit many of the elements of co-operation and conflict found in other animal societies. These social activities are extremely important to us: they determine our economic welfare; they produce a great deal of emotional turmoil, often providing the main reasons for whether we are happy or not; they determine how we are governed and how we structure our workplaces; and they even determine simple every day activities, such as how long we have to wait in queues.

Can some of the techniques used to study collective animal behavior be applied to understanding human societies? The answer is a qualified "yes." In narrowly defined social situations, such as in pedestrian movement and spectator crowds, some of the techniques used to understand collective animal behavior can be applied to humans. In wider situations, such as consumer decision-making and the "evolution" of fads and fashions, there could also be applications. Recent studies have looked at how our tendencies to buy particular items, find employment, and even commit crime change with the behaviors of those around us. Many of the underlying dynamics of these processes are similar to those seen in animal groups and this book seeks to highlight how these similarities arise.

Book Structure

Before exhibiting any form of collective behavior, animals must first come together. Chapter 2 looks at how and why animal groups form and the size distribution of these groups. This chapter sets the stage for more detailed investigations of the behavior of animals once they have formed groups.

One advantage of living in a group is information transfer. Chapter 3 investigates signals that social animals have evolved to share information about the presence and location of food, as well as cues that some individuals use to parasitize the information possessed by others. Chapter 4 shows that similar principles of information transfer underlie collective decision-making. During migration, cockroaches, ants, and bees all use similar rules to decide whether to move to a new home or shelter. These rules allow for consensus, whereby individuals "agree" where to move, and are thus able to choose the best of several options.

Some spectacular examples of collective behavior, which almost everyone has seen at some time, are moving animal groups. Chapter 5 looks at how simple models can capture the motion of fish schools, bird flocks, and insect swarms. Chapter 6 then turns to synchronization in time, looking at how fireflies co-ordinate their flashing and opera audiences co-ordinate their clapping.

Chapter 7 also concentrates on how simple rules can produce complex spatial patterns, but this time in the context of construction. This chapter looks at how termites and ants can build structures, such as nest mounds and trail networks, that are many orders of magnitude larger than they are. Chapter 8 looks at how interactions can lead to congestion and segregation. I discuss how economic and social systems often self-regulate to avoid these pitfalls, but sometimes also fail. Similar ideas can be applied to traffic congestion in ants and humans.

The repeated mantra that simple behavior by individuals can produce complex patterns can sometimes be taken too far. Animals are not just simple individual units; each possesses a great deal of behavioral and physiological complexity. In chapter 9, I describe some methods for dealing with such intrinsic complexity and modeling complicated networks of interactions.

Although throughout this book I try to emphasize both mechanistic and functional approaches to studying collective animal behavior, it is often the former approach that takes a larger role in each of the chapters. Chapter 10 readdresses this balance by returning to the examples elsewhere in the book and discussing how they might have evolved through natural selection. The main aim of this chapter is to show how a thorough mechanistic understanding can also clarify our ultimate functional understanding of biological systems.

Working with Models

Each chapter of this book has two or three boxes providing descriptions of key mathematical models. These models are most often simplified versions of those found elsewhere in the literature. The simplication and detailed description in the boxes has two complementary aims. Firstly, I hope that by simplifying the models I can illustrate better their central points. Secondly, I hope that the reader will be inspired to investigate and learn more about these models.

Understanding mathematical models requires more than just reading through their description. It is important to play around with them and find out how they work for different parameter values and react to small changes. For this reason, I have made the code for simulating most of the models presented in this book available online. The simulations run in Matlab and can be downloaded from www.collective-behavior.com.

The website also contains links to the homepages of many of the researchers mentioned in the book and links to key references.

— Chapter 2 —

Coming Together

Animal groups vary in size from two magpies sitting on a branch to plagues of millions of locusts crossing the desert. Not only do the sizes of groups vary between species, but they can change dramatically within species. In some cases, a change in group size depends on changes in the environment. For example, locust outbreaks are thought to originate where resources are patchily distributed, causing locusts to move towards these limited resources (Collett et al. 1998; Despland et al. 2004). In other cases, individuals in similar environments are found in very different-sized groups. Fishermen are used to such intrinsic variation in fish school size. Some days a net contains three fish, while the next day it contains tens of thousands (Bonabeau & Dagorn 1995). Human settlements also show similar variety in size, from tiny villages to massive cities, with differences in size arising without large differences in the environments in which they were originally founded (Reed 2001).

Can we then make general predictions about animal group sizes? In this chapter I approach the group size question from the two directions of functional and mechanistic explanation. The functional approach looks at how the costs and benefits of group membership can be used to calculate the optimal group size, at which individuals maximise their fitness, and the stable group size, at which no individual can improve its fitness by moving to another group. The mechanistic approach attempts to explain the large variation in group sizes observed empirically. By describing the mechanisms by which individuals join and leave groups a distribution of group sizes is predicted.

Optimal Group Size

There are many ways an individual can benefit from being a member of a group. The movement of a water skater as a predator approaches both confuses the predator and alerts other skaters of its presence (Treherne & Foster 1981); the starling in a flock can invest less time scanning for potential danger and more time probing the ground for food (Fernandez-Juricic et al. 2004b); the homing pigeon released with members of its roost can shorten its route home (Biro et al. 2006); the fish at the front of a school is less likely to be attacked than a straggler outside the group (Parrish 1989); and the pelican at the back of a v-formation saves energy in the wake of those in front (Weimerskirch et al. 2001). These and many other experimental observations explain why individuals form and join groups.

There are also always costs associated with group membership. While some less obvious costs, such as increased parasite burden (Brown & Brown 1986), have been demonstrated, they have not been studied empirically to the same degree as benefits (Krause & Ruxton 2002). In part this is because it is reasonable to assume that as group size increases, eventually so too does competition for local resources. For an overview and categorization of the different costs and benefits of group living see Krause & Ruxton (2002). For any single species, Brown & Brown's (1996) study of cliff swallows is probably the most comprehensive investigation of the costs and benefits of group living.

The functional approach to grouping considers how natural selection will act to shape group size. Individuals that live in groups where benefits outweigh costs will have a higher *fitness*, i.e., relative probability of survival and reproduction, than those in groups where the costs outweigh benefits. Thus a starting point for making predictions about how group sizes will evolve is to identify a *group size fitness function*. This fitness function can be calculated as the benefit minus the cost for individuals in groups of different size (figure 2.1).

The main practical consideration in determining the group size fitness function is finding a common currency or units, such as energy intake or time budgets, in which to measure costs and benefits (Krebs & Davies 1993). For example, Caraco used a theoretical model of the percentage of time yellow-eyed junkos feed, fight with each other, and scan for predators to make and test predictions about how behavior changes with group size (Caraco 1979a, 1979b) and how group size changes with food supply and predation risk (Caraco et al. 1980).

Whatever the currency chosen, a general observation about the group size function is that as groups become very large the costs will always exceed the benefits. Eventually local competition for resources outweighs any other benefits. The result of this observation is that the fitness function will have at least one maximum. Figure 2.2 gives three examples of theoretical group size fitness functions. In general, even if the group size has more than one local maximum, there is only one global maximum. This maximum is known as the *optimal group size*.

Determining the group size fitness function directly from the energy or time budget of individual animals for different group sizes is difficult in

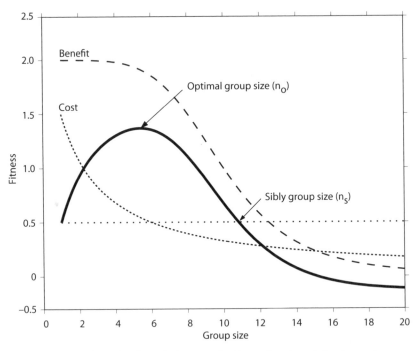

Figure 2.1 Derivation of a theoretical group size fitness function, $f(n)$, where n is group size. The thick dark line is the group size fitness function. This is derived from subtracting the cost function (dotted line) from the benefit function (dashed line). The cost function in this case is $3.75/(n + 1.5)$, which is a typical predation dilution curve: the rate at which an individual is attacked decreases in inverse proportion to the number of group members. We suppose the benefit function relates to rate of food intake, and is $2(1 - n^5/n^5 + K^5)$, where $K = 10$ is the group size at which individuals forage with exactly half the efficiency they forage with when alone. The optimal group size n_o is the value of n that gives the maximum difference between costs and benefits. The Sibly group size n_s is the maximum value of n for which $f(n) > f(1)$.

practice, not least because some unknown factor is easily omitted. There are however numerous empirical studies that have been able to relate group size to a particular variable that is likely to contribute to fitness. Pride (2005) found that stress, measured by levels of cortisol concentration, was higher for individuals in smaller and larger groups of Lemur. Individuals in intermediate sized groups showed lower stress (figure 2.3a). Brown and Brown (1996) found that during years where overall survival of young was low, cliff swallows in colonies of between 30 and 80 nests produced more surviving young than smaller or larger colonies. Due to difficulties in measuring survival of these offspring, they were unable to give a clear estimate of survival to adulthood. Without this estimate it is difficult to measure lifetime reproductive success, which accounts for the total number of individuals passed from one adult to the next generation of adults and is thus the preferable measure of fitness. Female lifetime reproductive success has been measured in social spiders (figure 2.3b) and individuals in intermediate sized groups of 23 to 107 had the highest fitness (Aviles & Tufino 1998).

Stable Group Size

While a particular group size may be optimal, this does not imply that it is stable. One theoretical prediction is that stable group sizes will usually be larger than the optimal group size. The argument for this prediction, first proposed by Sibly (1983), is that there is a benefit for individuals on their own or in smaller groups to join a group of optimal size, thus increasing the group size. More rigorously, this argument is made by considering a series of individuals arriving sequentially and choosing between a number of available resource sites. We assume their choices will be made on the basis of the fitness function, $f(n)$, in figure 2.2b. Further assuming there is no intrinsic difference between sites, the first arriving individual will choose a site at random. The second arriving individual will then choose the same site as the first, since it has a higher fitness there than on its own. Further individuals will continue to make the same decision provided the fitness gained from joining the group is larger than that of being on their own, $f(n + 1) > f(1)$. The important observation here is the advantage to the arriving individual to join a group even after that group has exceeded the optimal group size, n_o. If n_s is the largest group size for which $f(n_s) \geq f(1)$ then, under this process, all groups will become of size n_s. For most realistic fitness functions $n_s > n_o$ and the resulting group size will be larger than the optimal size (although see Giraldeau & Gillis (1985) for an exception to this rule where $n_s = n_o$).

The above argument has led some researchers to refer to n_s as the stable or equilibrium group size (Beauchamp & Fernandez-Juricic 2005; Clark & Mangel 1986; Giraldeau 2000). This interpretation suggests a paradox whereby groups reach a stable size for which membership confers no benefit over being alone, thus calling into question how grouping can evolve under free entry (Giraldeau 1988, 2000). The paradox arises however under three very strict, and in most cases biologically unrealistic, assumptions about how groups are formed: individuals (a) arrive sequentially starting with empty sites; (b) are unable to leave once they have chosen a site; and (c) are naïve to the order in which they arrive.

What happens if we relax assumptions (a) and (b) and individuals are free to move between sites? Box 2.A describes a simulation model, also

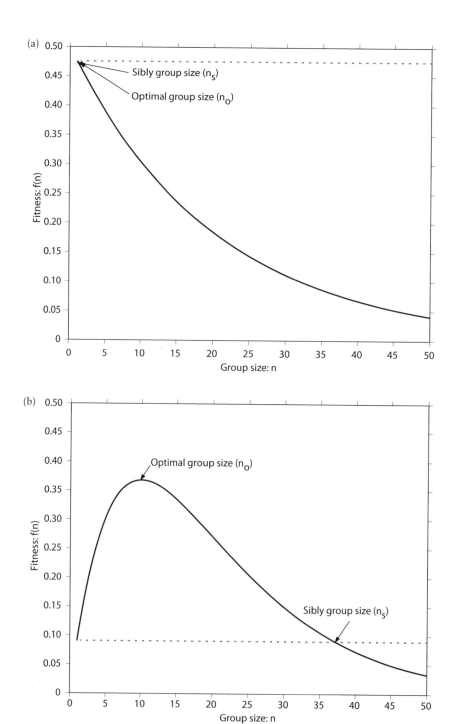

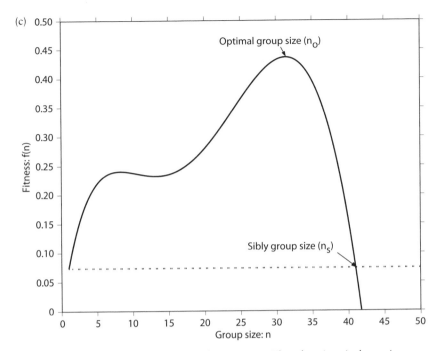

Figure 2.2 Theoretical group size fitness functions. (a) When there is a single maximum at a group size of one it is never advantageous for an individual to join a group; (b) a single maximum gives the optimal group size n_o, while the group size that has the same fitness as an individual on its own gives the Sibly group size n_s; (c) the fitness function has two local maxima but the global maximum is the optimal group size.

based on an argument first given by Sibly (1983), in which individuals are free to leave their current resource sites and join a site with higher fitness, with fitness being determined by the same function as in figure 2.2b. Figures 2.4a and 2.4b show the outcome of this model, given an initially random distribution of individuals between sites. Despite the highly variable starting distribution, the groups quickly converge to a stable size distribution with a mean slightly larger than the optimal group size. This stable group size distribution is not unique. Figures 2.4c and 2.4d show that if individuals are initially distributed with sizes close to n_o then the mean group size remains close to n_o.

In fact, most distributions of group sizes where all individuals are in groups of size greater than that which is optimal quickly become stable without greatly increasing in size. So unless the initial group size distribution has mean n_s there is no reason that it should be favored as the mean stable group size over any other mean group size greater than n_o. Indeed,

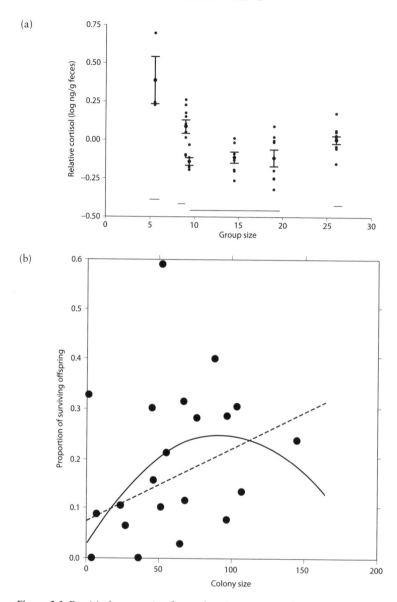

Figure 2.3 Empirical group size fitness functions. (a) Female cortisol levels for female ring-tailed lemurs averaged throughout the year per individual (reproduced from R. Ethan Pride, "Optimal Group Size and Seasonal Stress in Ring-Tailed Lemurs (Lemur catta)," *Behavioral Ecology*, April 2005, fig. 3, p. 555, by permission of Oxford University Press). (b) Proportion of surviving offspring per female in the colony for the social spider *Amelosimus eximius* (reproduced from Avilés & Tufiño, "Colony Size and Individual Fitness in the Social Spider Anelosimus eximius," *The American Naturalist*, January 1998, fig. 1, p. 409, © The University of Chicago Press).

Box 2.A. Sibly's Stable Group Size Model

Consider an environment with $s = 2000$ available sites. Assume initially that at half the sites, $i = 1$ to 1000, the number of individuals at the site, $n_i(0)$, is drawn from a uniform distribution with minimum $10 - r$ and maximum $10 + r$. Thus the average number of individuals at these occupied sites is 10 individuals, equal to the optimal group size in figure 2.2b. The other half of the sites, $i = 1001$ to 2000, are unoccupied, i.e., $n_i(0) = 0$. The unoccupied sites ensure that grouping in the model does not result simply from a limitation of available sites.

The rules of the model are as follows. On each time step t a random individual is picked. It then calculates the fitness function for all of the sites were it to move to that site, i.e., $f(n_j(t) + 1)$ for all sites apart from the site i that it is already at. In this case we use the group size fitness function shown in figure 2.2b, which is

$$f(n) = n \exp(n/10).$$

If $f(n_j(t) + 1) > f(n_i(t))$ for some j then the individual moves to the site that has the maximum value of $f(n_j(t) + 1)$. If more than one site has the same value of $f(n_j(t) + 1)$ then one of these sites is picked at random. This process is continued until no further moves are possible.

An example outcome of this process is shown in figure 2.4 for both wide ($r = 10$) and narrow ($r = 2$) initial distributions of group sizes. Small groups quickly reduce in size as members join larger groups. The optimal group size is unstable in both cases and is smaller than the stable group size. The stable group size differs with r, with larger stable group size for larger initial variation in distribution among sites. In no case is the stable group size as large as the Sibly group size, n_s.

the simulations suggest that for a wide range of initial group size distributions, stable group sizes will be only slightly larger than optimal.

There are of course many realistic situations in which individuals do arrive sequentially at a resource site and are unable to leave without incurring a cost. A typical example is birds arriving at a nesting site. However, before we predict stable group sizes close to n_s for sequential arrival we must consider what occurs if we remove assumption (c) and allow individuals to know how many individuals will arrive after them. In this

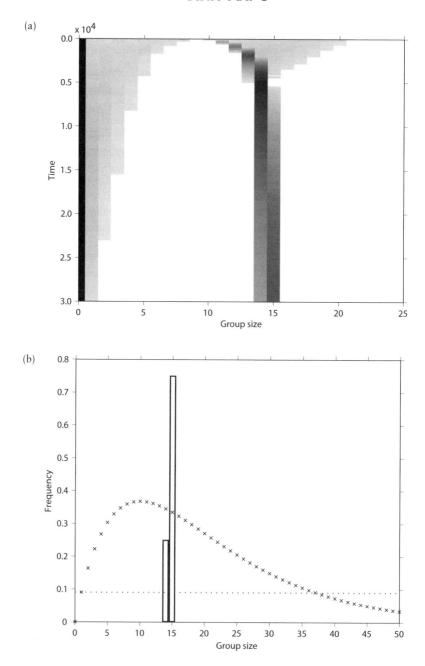

Figure 2.4 Outcome of Sibly's stable group size model for $r = 10$ (a, b) and $r = 2$ (c, d). (a) and (c) Time evolution of the group size distribution for 30,000 time steps. Shading indicates proportion of sites occupied by a particular number of individuals on a particular time

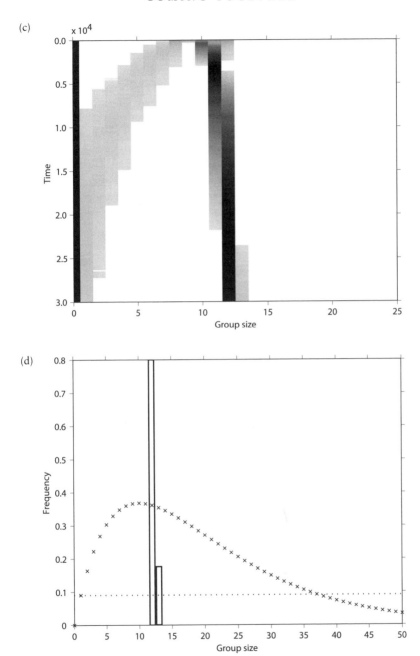

step. (b) and (d) Stable distribution of site occupation when no further moves are possible for the simulation. Crosses show group size fitness function and thin dotted line gives the Sibly group size, n_s.

case, it is best for early arrivals to occupy empty resource sites, secure in the knowledge that it will be best for later arrivals to join them. Given full knowledge of the sequence of arrivals it is conceivable that the stable strategy will result in group sizes very close to the optimal. A simple example of this can be constructed by considering four birds arriving with a group fitness function: $f(1) = 1$; $f(2) = 3$; $f(3) = 2$; and $f(4) = 1$. To optimize its fitness the third arrival must choose a site on its own (if the second has not already done so) thus ensuring that the fourth joins it. Turning the so-called group size paradox on its head, we see that even if some of the early arrivals are not joined, they will still have a fitness equal to that obtained if they ended up in a group of size n_s. Thus even with a high degree of error group sizes will in general be less than n_s. Although a complete knowledge of arrival sequence is not particularly realistic, changes in strategy dependent on arrival position are observed in birds (Brown & Brown 1996, chapter 13).

The above discussion highlights some of the difficulties in making general predictions about stable group sizes using the evolutionarily stable strategy models first proposed by Sibly. I would agree with the careful conclusion of Sibly (1983), "Flocks of optimal size are unstable and will tend to increase in size." However, group sizes only slightly above optimal are stable and only under a very limited set of assumptions is there a group size paradox. I thus follow the wording of Krause & Ruxton (2002) and call n_s the *Sibly group size*. The stable group size lies somewhere between the optimal, n_o, and the Sibly group size, n_s. Group size is likely to be highly dependent on the mechanisms through which groups form and the information available to potential group members about whether further individuals will join a group.

Natural Group Size Distributions

How do the actual sizes of animal groups compare to theoretical predictions about optimal and stable group sizes? Data to answer this question is lacking in many of the cases where group size fitness functions have been calculated, and where it is available it is often ambiguous (Krause & Ruxton 2002). One notable exception is Aviles & Tufino's (1998) study of social spiders. Figure 2.5a shows the distribution of group sizes of spider colonies under natural conditions. Compared to the predicted optimal group size of 50 (figure 2.3b) the mean group size is 425.6. Moreover, of the approximately 18,500 individual spiders surveyed, only 300 were in the optimal group size category of between 50 and 100 spiders. There is little evidence that the spiders usually obtain the optimal group size.

While these observations do provide support for the hypothesis that stable groups are larger than optimal, the most striking feature of the spider colony size distributions is that they are highly skewed. There are lots of small groups and a few exceptionally large groups. Similar group size distributions are seen throughout the animal kingdom. In addition to social spiders, figure 2.5b–d shows group size distributions for two mammalian herbivores and tuna fish schools. All these distributions have long tails corresponding to groups that are often several scales of magnitude larger than the modal group size.

Long-tailed group size distributions are clearly not expected from stable group size theory, which predicts a very narrow group size distribution (figure 2.4). This discrepancy between theory and data led Gerard et al. (2002) to question the validity of the stable group size approach to predicting group size. They suggested that although natural selection may play some yet to be established role in determining group size, the dynamics of fission and fusion in mobile mammalian and fish groups means that the sizes of the groups individuals find themselves in will vary widely, are seldom optimal, and certainly not stable. Aviles & Tufino (1998) are also sceptical about stable group size theory even for immobile spider aggregations. They cite population growth and dispersal costs as reasons for a wide range of group sizes. Although I would be less inclined than Gerard et al. to dismiss the optimal and stable group size approach entirely, it is clear from these empirical studies that a theory is needed that explains not only why groups of particular sizes arise, but also why there is such a variation in the size distribution of these groups.

Power Law Distributions

Long tailed distributions can often be described as a power law. A simple test of whether group size data might be power law distributed is to plot the logarithm of the group size against the logarithm of the frequency. If the data in this log-log plot is fitted by a straight line then it suggests that the data is power law distributed. Specifically, if the slope $-\alpha$ fits the data then a group size n occurs with frequency $p(n) \propto n^{-\alpha}$. Here α is referred to as the exponent of the power law.

Figure 2.5c shows such a log-log plot for American buffalo group sizes. This data fits a power law with exponent $\alpha = 1.04$. Figure 2.5d shows that the frequency of sizes of tuna fish catches is also fitted by a power law, with exponent $\alpha = 1.5$, over several orders of magnitude. Once group sizes become very large, frequency distributions usually tail off exponentially. The tuna fish data thus fits a truncated power law: a power law over several orders of magnitude but then tailing off exponentially

25

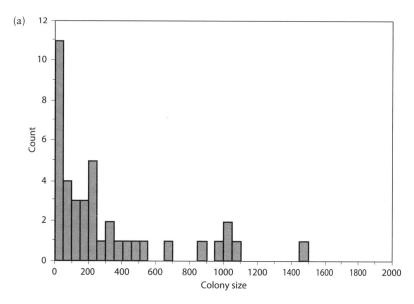

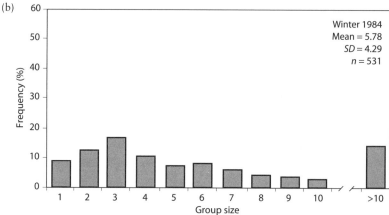

Figure 2.5 Group size distributions for (a) social spiders (reproduced from Avilés & Tufino, "Colony Size and Individual Fitness in the Social Spider Anelosimus eximius," *The American Naturalist*, January 1998, fig. 8, p. 409, © The University of Chicago Press); (b) roe deer in open cultivated planes with population density of 16–18 deer per ha (from J. F. Gerard et al., *Biological Bulletin*, 2002, fig. 1, 202: 275–285; reprinted with permission from the Marine Biological Laboratory, Woods Hole, MA); (c) American buffalo (Sinclair 1977). Here the data is plotted on a log-log scale. The solid line is the best linear regression to all points excluding the first point and gives log(frequency) = 8.76 − 1.04*log(group size). (d) Frequency of catches in terms of tonnes of Tuna fish caught in a net with a 2 km perimeter. (Reprinted with permission from E. Bonabeau & L. Dagorn 1995, "Possible Universality in the Size Distribution of Fish Schools," *Physical Review* E 51, fig. 1, R5220–R5223, © 1995 by the American Physical Society.) http://link.aps.org/doi/10.1103/PhysRevE.51.R5220. The fitted line has slope −3/2, giving a power law exponent of $\alpha = -3/2$.

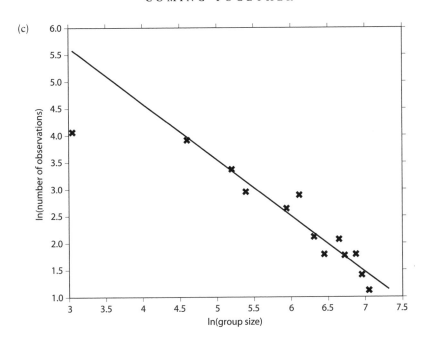

(c)

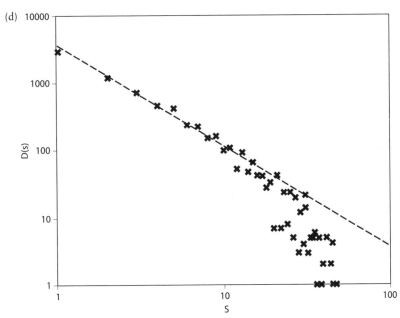

(d)

for very large groups. Since there is usually a limit to how large a real animal group can get, we expect most power laws to be truncated at some point. Truncated power laws give a reasonably good fit to many data of animal group sizes from spiders (Aviles & Tufino 1998), fish (Bonabeau et al. 1999; Niwa 1998; Niwa 2003), seals (Sjoberg et al. 2000), and mammalian herbivores (Gerard et al. 2002; Sinclair 1977).

Long tails in frequency distributions cause great excitement in the minds of theoretical physicists. These distributions are thought to characterize systems with highly non-linear dynamics or amplification of stochastic fluctuations (Sornette 2004). How are we meant to make biological sense of these ideas? We can start by investigating the assumptions underlying mathematical models that generate power laws. Do models that generate power laws have properties we can relate to the way individual animals interact? It turns out that there are a number of models, each based on reasonable biological assumptions that can generate power laws with slopes that match the data (Newman 2005; Sornette 2004). The problem is determining which is most realistic and could actually account for the observations.

Merge and Split Models

Power law distributions can be generated from very minimal assumptions about animal behavior. Bonabeau & Dagorn (1995) proposed a model for animal grouping based on a single assumption: that when groups meet they always merge to form a larger group. The model has s sites, each containing a group of size $n_i(t)$ at time step t. On each time step of the model each group picks a new site to visit at random. If two or more groups choose the same site then they merge, e.g., if the group at site i and the group at site j both move to site k, then $n_k(t+1) = n_i(t) + n_j(t)$. The resulting model is identical to a model of particles that stick together (Takayasu et al. 1988). When these particle (or animal) groups are equally likely to pick any of the available sites, and particles are added to the system at a constant rate, then the probability that a group is of size n is proportional to $n^{-3/2}$ (Takayasu 1989). This was very close to the power law exponent observed in the catch sizes of tuna (figure 2.5d).

Despite the claim that the above model might provide a universal law for fish school distributions, species other than tropical tuna do not have exponents of $-3/2$. Using computer simulations and further analytical results (Takayasu 1989; Takayasu et al. 1991), Bonabeau (1999) argued that exponents of between $-4/3$ and $-3/2$ could be accounted for by a reduction in the spatial dimension of the fishes' habitat; for example, attraction to specific resource sites. However, empirically measured exponents have a much wider range of between -0.7 and -1.8 (Bonabeau et al. 1999; Niwa 1998). A further limitation of the model is that it requires that individuals are continuously added, so that although the scaling rule continues to hold the population increases to infinity with time. If this assumption is removed then the theoretical exponent is -2 and, even less realistically, local populations at sites can become negative (Takayasu et al. 1991). There may be ways to overcome this technical limitation and recover an appropriate range of exponents, but these have not been fully investigated. In summary, while Bonabeau and Dagorn's work was useful in showing power laws in group size distributions, the theoretical model they used is not particularly biologically realistic nor a robust explanation of the available data.

Despite the limitations of early models, there does appear to be a universal scaling law for fish school sizes. Niwa (2003) took all available data on fish school sizes and re-plotted group sizes (N_i) versus frequency (W_i), this time dividing the group sizes by the expected group size experienced by an individual. This expected group size is given by

$$\langle N \rangle_P = \frac{\sum_{i=1}^{g} N_i^2 W_i}{\sum_{i=1}^{g} N_i W_i},$$

where g is the number of group size classes. $\langle N \rangle_P$ is not the same as the observed mean group size, which is $\sum_{i=1}^{g} N_i W_i$. Rather, $\langle N \rangle_P$ is the expected group size of an individual picked at random. $\langle N \rangle_P$ is always equal to or larger than the expected group size, since we are more likely to pick an individual in a larger group. Niwa found that by normalizing the data in this way, distributions for six different fish species all fall on the same curve (figure 2.6a). All these distributions had exponents close to -1 until normalized group size reaches one, at which point they tailed off exponentially.

The data is well fitted by the predictions of a simple model of group aggregation and breakup. The model's assumptions about aggregation were the same as Bonabeau and Dagorn's—groups move on each time step and when they meet they always merge—but Niwa further assumed that on each time step there is a fixed probability that groups break apart, splitting into two groups the size of which is uniformly distributed (see box 2.B for details of the model). The central prediction of this model is that the probability that a site contains a group of size N is

$$W(N) \propto N^{-1} \exp\left[-\frac{N}{<N>_P}\left(1 - \frac{e^{-N/<N>_P}}{2} \right) \right]. \qquad (2.1)$$

This equation captures the qualitative observation that group size distribution at first decreases inversely with N, but once the group size reaches $\langle N \rangle_p$ it starts to decrease exponentially. It fits both simulations of the above model (figure 2.7b) and the available fish data (figure 2.6a).

Niwa's work is remarkable in its generality. Bonabeau and Dagorn's model of truncated power laws has 4 parameters, all of which needed to be tuned for particular species. Niwa's model has one parameter that is naturally measured from the data and fits all available fish size data. In theory, measuring the average group size experienced by an individual allows the entire group size distribution to be predicted. Since equation 2.1 does not contain any model parameters, it is entirely independent of

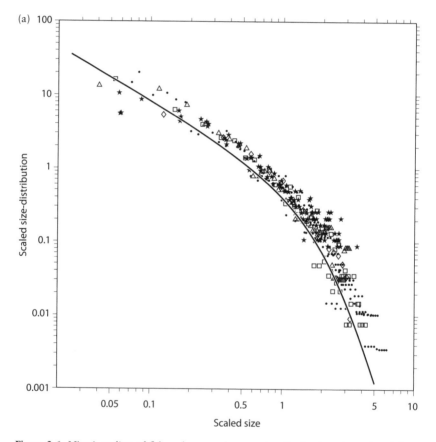

Figure 2.6. Niwa's scaling of fish and mammal group size distributions (reproduced from Hiro-Sato Niwa, "Power-Law versus Exponential Distributions of Animal Group Sizes," *Journal of Theoretical Biology*, October 2003, figs. 5 & 7, vol. 224 issue 4, 451–457, © Elsevier). (a) Empirical distribution of pelagic fish school sizes, six different species represented by

the rates at which groups merge and split. This may seem strange at first, but it should be borne in mind that $\langle N \rangle_p$ is determined by these rates. Indeed, Niwa (2004) showed that $\langle N \rangle_p \propto 1/p$ not only for his simple model, but also for a range of spatially explicit simulation models (see also chapter 5). Niwa has established a universal rule for fish schooling that does not depend on specific types of interactions and environmental structure. Provided fish schools merge when they meet and tend to split uniformly at random, we expect Niwa's predictions to hold.

The result does come with a couple of words of warning. When normalised so that they share at least one point in common and stretched out on a log-log plot, very different distributions can begin to appear

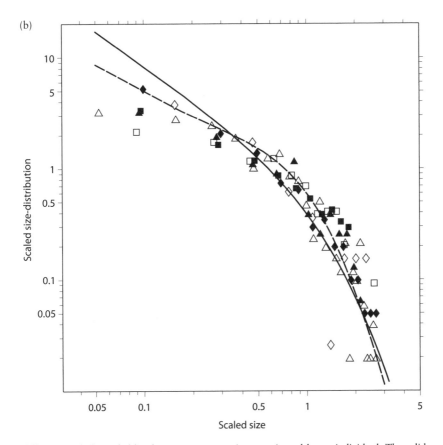

different symbols, scaled by the average group size experienced by an individual. The solid line is equation 2.1. (b) Empirical distribution of mammalian herbivore group sizes, six different species represented by different symbols. The solid line is equation 2.1 and the dotted line is a modified version of equation 2.1 with one extra parameter (see Niwa 2003 for details).

Box 2.B Niwa's Merge and Split Model

Assume that space is divided into s sites on which a total of m individuals are initially randomly distributed. The n_i individuals on site i are said to constitute a group. On each time step there are two stages to the model: move and split. First all groups move to a new site chosen uniformly at random. If two groups of size n_i and n_j meet at site k then they form a new group $n_k = n_i + n_j$, thus groups always merge when they meet. The same rule applies if three or more groups meet. After moving each group with a size greater than or equal to 2 will split into a pair of groups with probability p. When a group splits the size of the two components is chosen uniformly at random, so that all group sizes are equally likely. On the next time step the two split groups move separately to new randomly chosen sites, as do all unsplit groups, and the process continues. Figure 2.7a shows a time series of the number of individuals occupying a randomly chosen site for a simulation of this model for parameters $s = m = 2000$ and $p = 0.3$. Figure 2.7b shows the distribution generated by this simulation over 100,000 time steps.

Niwa used simulations to determine the form for the variance in the above model and applied results from Richmond (2001) to obtain an expression for the group size distribution in terms of $\langle N \rangle_P$. This expression is given by equation 2.1 in the main text. Niwa went on to show that for a variety of individual based models of schooling

$$\langle N \rangle_P \approx \frac{3.08\lambda\rho}{p\phi},$$

where λ is the probability per time step that a school merges; p is the probability per time step that a school splits; ρ is the population density, i.e., $\rho = \sum_{i=1}^{g} N_i W_i / s$; and ϕ is the proportion of the s sites occupied by a school (see chapter 5 box 5.A for details of spatially explicit models).

---

very similar. A similar method of data fitting has led to misleading conclusions about invariance in life history traits (Nee et al. 2005). Niwa's approach does not suffer from the same deficiencies, because group size and frequency are independent variables. The second warning is that the data used was based on fish catches and observations at fish aggregation devices. Such data is subject to sampling errors, with catches of certain sizes being preferred by the fishermen. With this in mind, it would be re-assuring to see a confirmation of these results for more fish species, with data collected using other measuring devices. I make these comments not because I doubt Niwa's findings but because, if the results were confirmed through independent field observations, his work would stand as one of the most fundamental laws of group behavior.

While Niwa's model might provide a universal rule for fish schooling, it does not appear to generalize to mammalian herbivores. Figure 2.6b shows herd size distributions for six different species. Although all six species lie on a similarly shaped curve, the data are not the same as given in equation 2.1. Niwa suggests that mammals might not break up according to the uniform splitting rule given in his model. Another possibility is that groups merge and split as a function of their size, and that the resulting group size distribution is a reflection of this behavior.

Preferential Attachment

With mammalian herbivores in mind, Gueron & Levin (1995) proposed a general framework for models where the probabilities of fission and fusion are a function of group size. They studied particular examples of this model in which the probability of two groups of size x and y merging could be written as $\psi(x,y) = \alpha a(x)a(y)$, while the probability of a group of size x splitting as $p(x) = \beta x a(x)$. They considered three cases: $a(x) = 1$; $a(x) = x$; and $a(x) = 1/x$. The use of $a(x)$ in both the splitting and joining probabilities produced a mathematical symmetry that allowed them to determine a function for group size frequency (Gueron 1998). Like Niwa's model they predict that the frequency of larger groups decreases exponentially with group size. This prediction was also made in Okubo's (1986) classic review of animal grouping, where he also argued that the available data on mammalian groups fitted an exponential model. However, closer examination of the data in figures 2.5b and 2.5c reveals that mammalian data has a longer tail than predicted by these models (see also Bonabeau et al. 1999). Although the framework of Gueron & Levin (1995) may well, for particular functional forms for $\psi(x,y)$ and $p(x)$, produce group size distributions similar to those seen in mammalian herbivores, the details of these forms have yet to be established.

One candidate for appropriate joining functions is preferential attachment. Preferential attachment is where the probability of an individual joining a group increases with group size. Box 2.C gives an example of a model where the probability of an individual joining a group is a linearly increasing function of its size, while the probability of a group splitting is

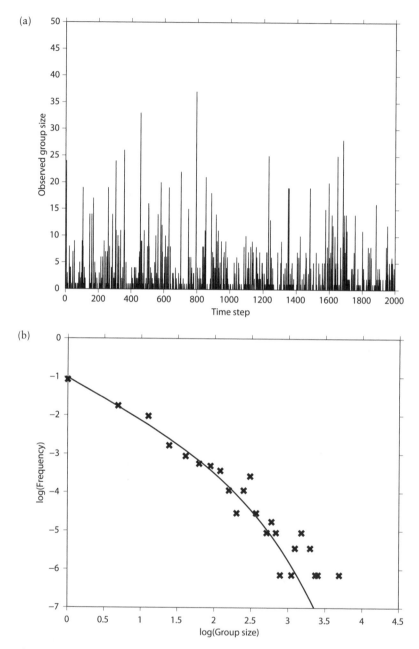

Figure 2.7. Niwa's merge and split model (a and b) and a preferential attachment model (c and d). (a) A time series of the number of individuals occupying a randomly chosen site for a simulation of Niwa's model (see box 2.B for details) for parameters $s = m = 2000$ and $p = 0.3$. (b) The group size distribution generated by this simulation over 100,000 time

(c)

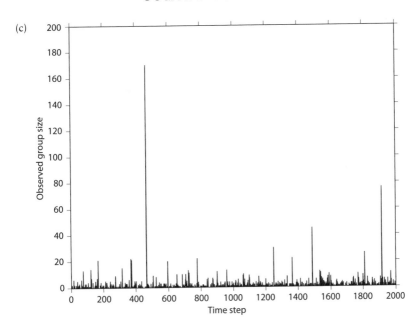

(d)

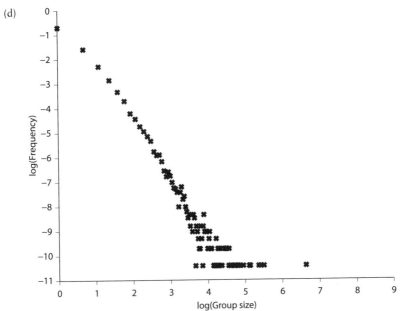

steps. (c) A time series of the number of individuals occupying a randomly chosen site for a simulation of a preferential attachment model (see box 2.C for details) for parameters $s = m = 2000$ and $c = 1$. (d) The group size distribution generated by this simulation over 100,000 time steps.

Box 2.C Preferential Attachment Model

Price (1976) proposed a preferential attachment model for scientific citations. In the model, papers are written one after another with no overlap or delay in publication time and each paper cites b previous papers. When each new paper is published, the probability that a currently existing paper i is cited by this new paper is proportional to the number of times, n_i, that the existing has already been cited. In particular, the probability it is cited is

$$\frac{(n_i + c)}{\sum_{j=1}^{m}(n_j + c)} \qquad (2.C.1)$$

where m is the total number of papers and c is a constant. This model is known as preferential attachment since the probability of attachment increases with the number of previous attachments (i.e., citations).

We let p_n be the probability that under this system a paper is cited n times. Newman (2005), following a method first developed by Simon (1955), shows that the tail of the distribution of citations is according to a power law, i.e., for large n, $p_n \sim n^\alpha$, with

$$\alpha = 2 + \frac{c}{b}. \qquad (2.C.2)$$

Empirically, we see that paper citations are distributed with a power law with slope 3.04 (figure 2.8b). Were the model to fit the data we would thus predict that $c \approx b$. In general, appropriate choice of c and b can produce a power law with any slope greater than 2.

The above model applies in cases where the population continues to grow, as it clearly does with scientific papers. In modeling animal populations that do not change in total population, as in Niwa's model, that space is divided into s sites on which a total of m individuals are initially randomly distributed. The n_i individuals on site i are said to constitute a group. In the spirit of preferential attachment, on each time step we choose a site i at random and remove all individuals, modeling perhaps a disturbance by a predator. We then redistribute them between the sites according to equation 2.C.1. Figure 2.7c shows a time series of the size of the group at the randomly chosen site and figure 2.7d shows the distribution of these group sizes on a log-log plot. This simulation also appears to give a power law distribution.

independent of group size. This particular model generates a power law distribution of group sizes with an exponent of approximately 2.5. In this model I assumed that the population size remains constant. This assumption is consistent with the dynamics of mobile animal groups where total population size usually changes more slowly than the rate at which individuals leave or join groups. While the mathematical properties of constant population models have not been extensively investigated, analysis of preferential attachment models in which the population continues to grow suggests that, depending on the details of the rules for attachment, power laws with exponents of greater than or equal to two can be generated (again see box 2.C).

Many of the distributions associated with human behavior exhibit power laws with exponents greater than or equal to two. A now classic example is the growth and connection of websites on the World Wide Web. The frequency of the number of connections to websites follows a power law with a slope of 2.1 over four orders of magnitude (Barabasi & Albert 1999; Barabasi et al. 1999). Figure 2.8 shows a large number of examples of distributions that have been claimed to follow power laws. Newman (2005), who produced this figure, emphasizes that it is difficult to confirm that these data really do follow a single power law rather than multiply-overlaid power law or non-power law distributions. Furthermore, since power laws often only hold in the tail of a distribution, a somewhat arbitrary cut-off point has to be selected above which the exponent α is estimated. These technical limitations do not substantially detract from the ubiquity of power laws (Ball 2004; Buchanan 2000). Across many different types of systems, not only those associated with humans but also in the physical and biological world, power laws provide a good fit to the distribution of events occurring in these systems.

Is the ubiquity of power laws a consequence of preferential attachment mechanisms? The key question is whether the individuals that contribute to systems with power law distributions behave in a way consistent with preferential attachment. We can see how this could be the case with, for example, the growth of the internet or citations of scientific papers, where the probability that we link to a web site or cite a particular paper increases with the number of previous links or citations.

Let's consider preferential attachment applied to scientific citations. Assuming that the constant c in equation 2.C.1 is equal to the mean number of citations per paper then using equation 2.C.2 we recover the power law with exponent approximately 3 seen in the science citation data (figure 2.8). In the model, initial citations are chosen entirely at random then further citations are made according to how many previous citations have been made, rather than on the basis of anything written in the papers.

We are led to the rather disturbing conclusion that citations may be due entirely to amplifications of initially random decisions on the part of scientists and are independent of the supposed quality of the papers.

While not ruling out the above model of scientific citations it should be pointed out that it is by no means a unique explanation. For example, although the famous bell-shaped or Normal curve is an accurate description of the empirical distribution of IQ near to the mean of the distribution, the tails of this distribution are much wider than predicted by the Normal distribution (Burt 1963). In general, large deviations in distributions are often better characterized by power laws than the Normal approximation (Sornette 2004). Assume that the quality of papers is proportional to author IQ, and scientists working in academia come from the upper tail of the IQ distribution (I do realise the limitations of this assumption). If papers are cited in proportion to their quality then the distribution of citations will simply reflect the power law distribution

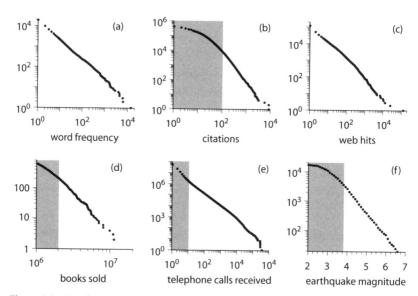

Figure 2.8. Distributions or "rank/frequency plots" of twelve quantities reputed to follow power laws (reproduced from M. E. J. Newman, "Power laws, Pareto distributions, and Zipf's law," *Contemporary Physics* vol. 46, no. 5, September–October 2005, 323–351, fig. 4). Data in the shaded regions were excluded from the calculation of the estimated power law exponents, α. (a) Numbers of occurrences of words in the novel Moby Dick by Hermann Melville, $\alpha = 2.20$. (b) Numbers of citations to scientific papers published in 1981, from time of publication until June 1997, $\alpha = 3.04$. (c) Numbers of hits on web sites by 60,000 users of the America Online Internet service for the day of 01 December 1997, $\alpha = 2.40$. (d) Numbers of copies of bestselling books sold in the US between 1895 and 1965, $\alpha = 3.51$. (e) Number of calls received by AT&T telephone customers in the US for a

of the IQ of their authors. Likewise, the links to webpages might be proportional to the intelligence of their designer or the funds possessed by their owner (figure 2.8j).

Neither preferential attachment nor extreme IQs provide entirely satisfactory mechanistic explanations of the power laws arising in figure 2.8. Indeed, there are at least a dozen distinct mathematical models—from self-organised criticality (Bak 1996) to highly optimized tolerance (Carlson & Doyle 2002; Doyle & Carlson 2000)—in which power laws can be derived (Newman 2005; Sornette 2004). Even the causes of power law distributions in physical systems such as meteorite sizes and earthquakes have no generally accepted explanations. In themselves, power laws provide a very weak predictor of the mechanisms that generate them. We should not however be overly discouraged by these observations. Each of the mechanisms for generating power laws has its own set

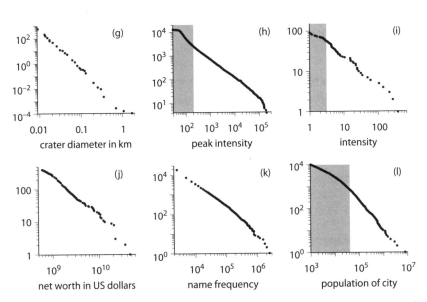

single day, $\alpha = 2.22$. (f) Magnitude of earthquakes in California between January 1910 and May 1992. Magnitude is proportional to the logarithm of the maximum amplitude of the earthquake, and hence the distribution obeys a power law even though the horizontal axis is linear, $\alpha = 3.04$. (g) Diameter of craters on the moon. Vertical axis is measured per km, $\alpha = 3.14$ (h) Peak gamma-ray intensity of solar flares in counts per second, measured from Earth orbit between February 1980 and November 1989, $\alpha = 1.83$. (i) Intensity of wars from 1816 to 1980, measured as battle deaths per 10,000 of the population of the participating countries, $\alpha = 1.80$. (j) Aggregate net worth in dollars of the richest individuals in the US in October 2003, $\alpha = 2.09$. (k) Frequency of occurrence of family names in the US in the year 1990, $\alpha = 1.94$. (l) Populations of US cities in the year 2000, $\alpha = 2.30$.

of assumptions, which are experimentally testable. Further experiments can be performed to test the various models against each other.

Bearing in mind our general caution about power laws, we can now begin to think about how to apply our models to explaining the group size distribution of, for example, mammalian herbivores. The example model I give in box 2.C probably does not encompass the behavioral rules whereby buffalo groups join and split. Furthermore, the model also gives a significantly larger exponent than $\alpha \approx 1.04$ estimated from the data in figure 2.5c. However, the preferential attachment model incorporates realistic behavioral rules into grouping models: individuals prefer to join larger groups, which are then split by random disturbance. With further refinement, this model may begin to capture empirically measured behavior of real animals.

The possibility of including behavioral rules whereby individuals attempt to maximize some variable, in this case group size, brings me back to the functional models with which I began this chapter. Power law distributed group sizes and the instability of the optimal group size become complementary ideas. Preferential attachment is the mechanism by which individuals are more likely to attach themselves to larger groups. The functional reason for this strategy follows from the advantage of being in a larger group, even if that group is larger than the optimal size.

Group Size and Population Density

Niwa's and Sibly's models give different predictions about how group size changes with population density. While keeping the same basic rules for merging and splitting, Niwa (2004) showed that, for a variety of individual based models of schooling, the mean group size experienced by an individual, $\langle N \rangle_P$, was proportional to the population density. Thus Niwa's model predicts that mean group size will strictly increase with population density. Sibly's stable group size model (box 2.A) predicts that, provided the total population is larger than the Sibly group size, group size will remain constant as population density increases. Under this model, increases in population density will lead to further groups being created, of stable group size somewhere between the optimal and Sibly group size. Under Niwa's model we also expect increases in group number with population density, but this would be less pronounced than under Sibly's model.

Laboratory experiments on killifish support the predictions of Niwa's model (Hensor et al. 2005). Both group size and group number increased with population density. There was no indication of the modal group size leveling off at a particular "stable" number and it appears that the distribution of group sizes had a large variance. Hensor et al. (2005) developed an individual-based model, based on mechanistic principles of local individual attraction, which gave a very good match to the experimental data.

Field experiments on killifish gave qualitatively similar results to the laboratory experiments (Hensor et al. 2005). Both group number and group size increased with population density. Quantitatively, however, results from laboratory and field were very different. The number of groups was much smaller and group sizes were much larger in the field than in the laboratory, and were no longer consistent with Hensor et al.'s model. The differences between laboratory experiments and fieldwork may be accounted for in terms of environmental heterogeneity. The fish may be attracted to a certain feature of their environment that simply is not present in homogeneous laboratory conditions. Furthermore, Hensor et al. found that fish body size has an important role in determining group size distribution. The failure of models to accurately predict the outcome of field experiments brings me to a final word of warning about the assumptions that underlie the models discussed in this chapter.

Alternative Explanations for Grouping

Most of the models discussed in this chapter assume that groups consist of genetically unrelated individuals that have similarly shaped group size fitness functions and live in relatively homogeneous environments. One species for which these assumptions have been explicitly tested are cliff swallows, which are not genetically related, exhibit no relationship between site availability and group size, but do have between-individual differences in group size fitness functions (Brown & Brown 1996). Indeed, Brown & Brown attribute these last differences to much of the between-group size variation observed in cliff swallows. In general we can't hope that these assumptions hold exactly for all the species we are interested in but we can expect them to be a reasonable approximation of reality.

Particular care should be taken with the assumption of environmental homogeneity. Figure 2.8 shows that many features of the physical world have distributions similar to those seen in animal groups. The sizes of animal groups could then simply be attraction to particular physical features, rather than aggregation in response to other animals. Another possibility is that the distribution of a predator species is simply a reflection of the distribution of prey. For example, a predatory fish might gain greatest fitness foraging alone but due to the clustered distribution of its prey it is found in group size distributions similar to that of its prey.

Giraldeau & Caraco (2000) refer to this type of attraction to resources as a "dispersion economy" (group size fitness function as in figure 2.2a) while attraction to conspecifics is referred to as an "aggregation economy" (group size fitness function as in figure 2.2b,c). It is usually straightforward to discern if animals are part of a dispersion economy by testing whether individuals in homogeneous environments are attracted or repelled by conspecifics. More difficult is separating effects of attraction to aspects of the environment from those to other individuals in aggregation economies. If an animal is weakly attracted to a particular environmental feature then this weak attraction can be amplified as others copy the choices made by some individuals. One experimental approach uses binary choice tests where individuals are presented with two identical environments (Amé et al. 2004; Goss et al. 1989). I will discuss such tests in more detail in the next chapter.

Linking Mechanistic and Functional Approaches

There is less contradiction between mechanistic models discussed in the second half of this chapter and the functional models than is sometimes supposed. All mechanistic models make implicit assumptions about the group size preferred by individuals in groups. The rules of the models mean that individuals experience a typical group size. For example, in Niwa's model the group size experienced by an individual is $\langle N \rangle_p \propto 1/p$ and this can be controlled by the individuals by adjusting the rate at which they split, i.e., changing p. What is not investigated in these models is how an individual can adjust its probability of leaving a group in order to increase its own fitness. Indeed, it is usually the probability of a group joining another group that is used in these models, rather than the probability of an individual leaving or joining a group.

Surprisingly, no one has investigated fission and fusion models within the context of optimizing group size. This is unfortunate since basic fission and fusion or joining and leaving rates can be empirically measured and these models could be used to make predictions about what animals are trying to optimize. The approach of Gueron & Levin (1995) would be a good starting point, but the symmetrical fusion and fission used in their model is counter-intuitive. For example, in their model large groups are simultaneously more likely to split and to join other groups. These assumptions are particularly strange in the light of optimal group size theory, where we might expect merging to increase below optimal group size and splitting to increase above optimal group size, and vice versa.

An interesting question is the circumstances under which individuals following a simple set of leaving and joining rules will reach a group size distribution with a mean or mode close to the optimal group size. Beauchamp & Fernandez-Juricic (2005) have made a start on this question. They assumed that individuals decide to leave resource sites on the basis of an estimate of their food intake at that site (Bernstein et al. 1988). The food intake is then a function that first increases but later decreases with group size (e.g., figure 2.2b). Using this model they showed that groups formed with a modal size near to that of the optimal group size and much lower than the Sibly group size. Furthermore, the distribution of group sizes had a large variance consistent with empirical data.

More work is needed in understanding how and why groups of unrelated individuals form. Indeed, it is quite surprising how little this basic problem of collective behavior has been studied either theoretically or experimentally. In comparison to aspects of how individuals act once established in groups, the process by which they have formed has received less attention. This disparity may be due to the fact that without understanding aspects such as information transfer, decision-making, and synchronization we cannot discern the benefits and costs of grouping. The models presented in this chapter, and particularly the work of Niwa, should however encourage us that it is possible to make predictions about individuals coming together without knowing the details of what animals do once the group has formed.

— Chapter 3 —

Information Transfer

A key benefit of being near to others is access to information. Animals often live in environments where resources are distributed in patches that exist only temporarily. In such an environment, a single individual has a very low rate of finding a resource patch if they search independently. When large numbers of individuals search at the same time, however, the probability that one of them finds one of the patches is considerably larger. If individuals are able to monitor and use the discoveries of others in their own search, they can increase their own rate of finding resources.

Many of the mechanisms underlying information transfer are the same across species. Underlying all information transfer is some form of positive feedback: one individual finds food, a second moves towards the first individual and then still a third moves towards the second and so on. This chapter uses a couple of simple mathematical models of positive feedback to provide a reference point for different forms of information transfer. These models help us classify information transfer seen across species.

Information Centers

Living in a communal nest or den provides a good opportunity for information transfer, and in some cases may be the reason communal living has evolved (Zahavi 1971). Individuals returning to the nest with food also carry with them information about its location and quality. This information can be used by nestmates. In social insects, sophisticated signals have evolved to actively communicate food discoveries. Such signals have also evolved in some birds and mammals, but they are not a necessary requirement for information transfer. Communally nesting animals can also use passive cues, such as flight direction and smell, to identify where food has come from.

Ant Pheromone Trails

Many species of ants deposit chemicals, known as pheromones, to mark the route from food to nest (Hölldobler & Wilson 1990; Wilson 1971). After finding a food source and feeding, an ant returns to the nest, pausing at regular intervals on its way to leave small amounts of pheromone. The ant then makes repeated trips from nest to food source, often leaving more pheromone to reinforce its trail. Other ants, which were previously unaware of the food source but encounter the trail, follow the trail and find the food. Once they have collected food, these follower ants also leave pheromone on their return journey. Through this reinforcement, the pheromone trail builds up and after a short time we see a steady trail of ants walking between food and nest. Pheromone trails are formed purely on the basis of local information. They are started by a single individual or a small group of ants responding to the presence of food and they are reinforced by ants that encounter and follow the trail.

Pheromone trails act not only to inform nestmates where food is located, but can also be used to find the shortest path to it. For example, Beckers et al. (1992b) presented starved colonies of the ant *Lasius niger* with two alternative bridges between food and nest, then measured the number of ants using the two bridges 30 minutes after the first ant had found food. When one of the bridges was only 40% longer than the other, over 80% of the ants took the shorter bridge in 16 out of the 20 experimental trials (figure 3.1a). Individual ants make little or no comparison of the two bridges, instead the slightly longer trip time means that pheromone is laid less rapidly on the longer bridge. Thus when trail following ants make the choice between two bridges they detect a higher concentration of pheromone on one of the bridges, i.e., the shorter one (Beckers et al. 1993). As a result the shorter bridge is chosen with a higher probability by the follower ants and when these ants return home they further reinforce the shortest path. Since pheromone continually evaporates on both paths but is more strongly reinforced only on the shortest path, the ants rapidly concentrate their trail on the shorter path.

The basic principle underlying pheromone trails is positive feedback: the one ant that first finds the food starts a feedback loop as more and more ants are recruited, and as more ants are recruited the rate of recruitment increases further. Positive feedback can be succinctly captured by a differential equation model (box 3.A). The key assumption in the model is that the probability that an ant takes bridge X is proportional to

$$\frac{(x+k)^{\alpha}}{(x+k)^{\alpha}+(y+k)^{\alpha}}, \tag{3.1}$$

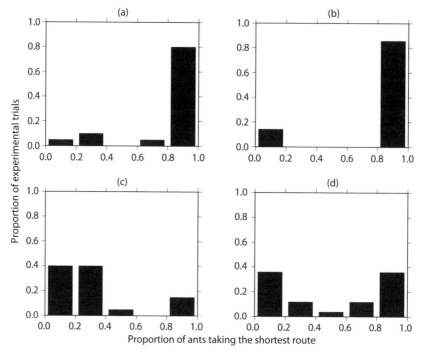

Figure 3.1. Outcome of Beckers' experiment with ants on bridges. Each panel gives the distribution of the proportion of experiments in which the ants followed the shortest available path to food. The length ratios of the paths were: (a) 1:1.4; (b) 1:2; (c) 1:2, but with a second path introduced later; and (d) 1:1. Results reproduced from Beckers et al. (1992b) for (a),(b), and (d), and Camazine et al. (2001) for (c).

where x and y are the amount of pheromone on each of the respective bridges, and k and α are constants. k and α have been measured for *Lasius niger* (Beckers et al. 1993), in which case $\alpha = 2$, and argentine ants (Vittori et al. 2006), where $\alpha = 5$.

The fact that $\alpha > 1$ means that the ants' response to pheromone is non-linear and differences between the amount of pheromone on two alternative bridges are amplified. Solving the model shows how a relatively small bias in the travel time is amplified to give a large bias in number of ants taking a particular route (figure 3.2a). The model predicts that at equilibrium the bias to the shorter route will be much greater than a simple ratio of route lengths. This prediction is reflected in the data in figures 3.1a and 3.1b where small differences in route lengths are amplified so that nearly all the ants take the shortest route in the majority of experiments.

The model makes a further counter-intuitive prediction: that a small initial bias towards the longer of the two routes can be amplified so that

Box 3.A Model of Ant Foraging and Symmetry Breaking

Deneubourg et al. (1990a), and later Beckers et al. (1993), developed the following model to describe the trail-laying of ants that are offered a binary choice between two alternative bridges between their nest and food. Each ant approaching the branching point will choose bridge X with probability

$$\frac{(x+k)^\alpha}{(x+k)^\alpha + (y+k)^\alpha}, \tag{3.A.1}$$

where x, respectively y, is the concentration of pheromone on bridge X respectively Y; k and α are constants. Beckers et al. (1993) determined experimentally that for *Lasius niger* ants, measured $k = 6$ and $\alpha = 2$. They further assumed: that ants left the nest at a constant rate ϕ; that individual ants would take the same bridge back on their outward and return journeys; that these ants deposit an amount of pheromone, q_x or q_y, in proportion to the quality of the food or the length of the bridge; and that this pheromone will evaporate at a rate v (see Camazine et al. 2001 page 232 for a detailed list of assumptions). Under these assumptions a differential equation model can be written to express the rate of change of the pheromone concentration, or equivalently the number of ants on the two bridges:

$$\frac{dx}{dt} = \phi q_x \frac{(x+k)^\alpha}{(x+k)^\alpha + (y+k)^\alpha} - vx \tag{3.A.2}$$

$$\frac{dy}{dt} = \phi q_y \frac{(y+k)^\alpha}{(x+k)^\alpha + (y+k)^\alpha} - vy \tag{3.A.3}$$

(Nicolis & Deneubourg 1999). See Sumpter & Pratt (2003) for details of how to derive this and similar differential equation models.

Figures 3.2a and 3.2b show numerical solutions of equations (3.A.2) and (3.A.3) through time for different initial conditions. For these parameter values, the bridge ultimately taken is determined by the initial conditions. If the majority initially take the shortest bridge (X) then this bridge is ultimately chosen (figure

(Box 3.A continued on next page)

3.2a), but if a majority take the longer bridge (Y) in the beginning then it is instead chosen (figure 3.2b). Figure 3.2c shows how the bridge ultimately chosen depends upon the initial conditions and the flow rate ϕ out of the colony. This figure is known as a bifurcation diagram. For low flow rates there is a single stable equilibrium where the ants divide themselves between the two bridges roughly in proportion to their quality. For high flow rates there are two stable equilibriums, one corresponding to the majority of ants using bridge X, the other to the majority of ants using bridge Y.

In the case where the bridges are equal, $q = q_x = q_y$ we can analyze these equations to find the equilibriums. An equilibrium occurs when $dx/dt = dy/dt = 0$, i.e.,

$$\frac{\phi q}{v}(x + k)^{\alpha} = x\left((x + k)^{\alpha} + (y + k)^{\alpha}\right)$$

$$\frac{\phi q}{v}(y + k)^{\alpha} = y\left((x + k)^{\alpha} + (y + k)^{\alpha}\right)$$

Adding the left and right hand sides of both equations shows that $x + y = \phi q/v$. Dividing the first equation by the second then gives

$$\left(\frac{\phi q}{v} - x\right)(x + k)^{\alpha} = x\left(\frac{\phi q}{v} - x + k\right)^{\alpha} \tag{3.A.3}$$

In the case where $\alpha = 2$ there are equilibriums at $x = y = \phi q/2v$ and

$$x = \frac{\phi q}{2v} + \sqrt{\left(\frac{\phi q}{2v}\right)^2 - k^2}, y = \frac{\phi q}{2v} - x.$$

The first equilibrium only exists when $\phi q/2v > k$, thus in terms of flow out of the nest $\phi = 2kv/q$ is a bifurcation point at which the number of stable equilibriums changes. Such a bifurcation is known as *symmetry breaking* since above this flow rate there exist two different stable equilibriums, one of which corresponds to more ants using bridge X and the other to more ants using bridge Y. Symmetry is broken despite the fact that the recruitment function to both bridges is the same. For further analysis of this model, see Nicolis & Deneubourg (1999). A typical simulation and a bifurcation diagram for this model when $q = q_x = q_y$ are shown in figure 3.3.

Symmetry breaking requires that $\alpha > 1$, i.e., that the probability of taking a particular bridge is disproportionately higher for the one with more pheromone on it. In the text I discuss a model of honeybee dance following that has the same functional form as the bridge following, but $\alpha = 1$. Here the probability is of following a dance to a particular flower patch, rather than choosing a particular bridge, but the principle remains the same. Substituting $\alpha = 1$ into equation 3.A.4 gives only a single stable equilibrium $x = y = \phi q/2v$, independent of the flow of bees or the initial number of bees at the two flower patches. There is thus no symmetry breaking in honeybee dance recruitment. Even when the quality of the two feeders is different there is only a single equilibrium (see figure 3.4).

this route will ultimately be taken by the majority of the ants (figure 3.2b). This prediction is borne out by the experimental data, in which the three trials that were not biased to the shorter route were biased toward the longer route (figures 3.1a and 3.1b). To further test how initial conditions determined final outcome, Beckers repeated the bridge experiments, this time starting with only the longer bridge available (28 cm between nest and food). Once the ants established a trail on this bridge, a shorter bridge (14 cm) was introduced. The established feedback on the longer bridge was so strong that in 16 out of 20 trials the ants did not switch to the shorter bridge (figure 3.1c). Strong positive feedback locked the ants in a suboptimal path choice.

Positive feedback with non-linear responses to pheromone differences also amplifies small environmental differences. Dussutour et al. (2005a) offered colonies of *Lasius niger* ants two equal length bridges to food, one with a 2-mm high wall on the inner edge of the bridge. In 16 out of 19 trials (84%) the majority of the ants followed the bridge with the wall. In itself this result is not surprising, since it is well known that ants like many other animals follow edges. Interestingly, however, when Dussutour repeated the experiment, this time only letting one ant at a time onto the bridges, she found that 66% of ants chose the bridge with the wall. Thus, while individual ants did show a small bias towards wall following, the strong tendency of ant trails to follow walls is due to an amplification of this small initial bias.

The degree to which an initial bias is amplified and the ants show a strong preference for a particular route depends upon the rate at which the ants leave the colony. Figure 3.2c shows a *bifurcation diagram* of how the equilibrium number of ants taking each route depends upon the initial

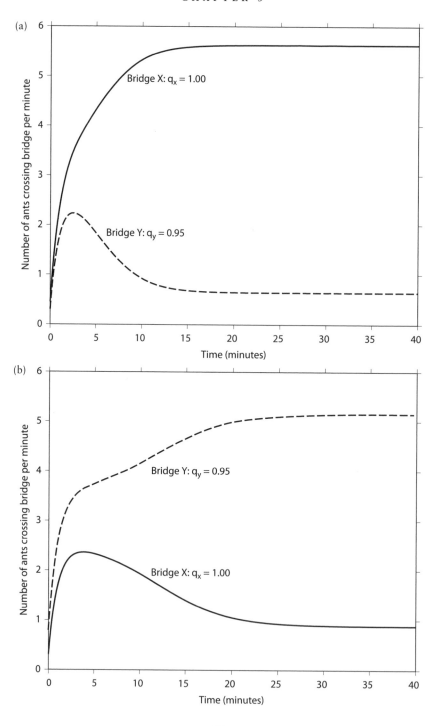

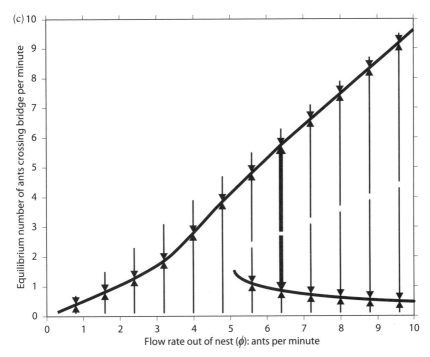

Figure 3.2. Outcome of model in box 3.A when bridge lengths are unequal: (a) Simulation of model where the initial bias is to bridge X, $x(0) = 0.35$ and $y(0) = 0.2$. (b) Simulation of model where the initial bias is to route 2, $x(0) = 0.3$ and $y(0) = 0.1$. (c) Bifurcation diagram for model. Thick lines going from left to right are the stable equilibrium, while the arrows show which set of initial conditions arrive at different equilibriums. The thicker arrow in (c) denotes the case where $\phi = 6.3$, corresponding to the simulations in (a) and (b). Other parameter values are $q_x = 1$, $q_y = 0.95$, $v = 1$, $k = 2$ and $\alpha = 2$.

number taking bridge X and the flow of ants out of the nest. When the flow out of the nest is low there is a unique equilibrium whereby only slightly more ants take the shortest route and the longer route continues to be used. At a flow rate of about 3.5 ants per minute the bias towards taking bridge X (the shortest route) increases, and once 5 ants per minute are flowing out of the nest, nearly all of them take bridge X. However, at flows of just over 5 ants per minute, a second stable equilibrium appears, i.e., a bifurcation occurs. For these flows, an initial bias towards bridge Y will be preserved and the majority of ants will take bridge Y instead of the shorter bridge X.

Strong preference at the level of the group can occur even in the absence of any bias in the length of the bridges. The model in box 3.A predicts that given two equal routes to the feeder, instead of splitting 50:50 between the two routes one route will be chosen over the other (figure 3.3a). When

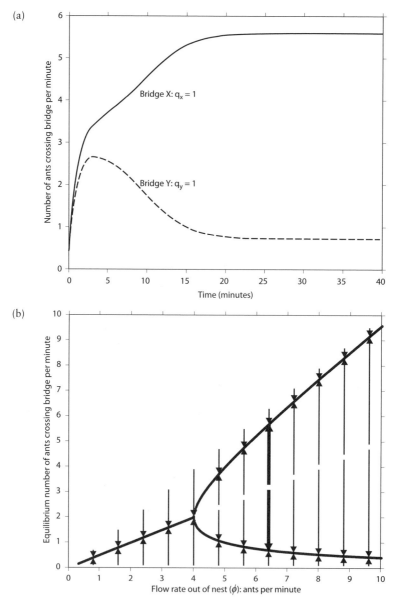

Figure 3.3. Outcome of model in box 3.A when paths lengths are equal. (a) Simulation of model where the initial bias is to bridge X, $x(0) = 0.35$ and $y(0) = 0.2$. Swapping these initial conditions would create an equilibrium, where more ants use bridge Y. (b) Bifurcation diagram for model. Thick lines going from left to right are the stable equilibrium, while the arrows show how different initial conditions go to different equilibriums. The thicker arrow in b denotes the case where $\phi = 6.3$, corresponding to the simulation in (a). Other parameter values are $q_x = q_y = 1$, $v = 1$, $k = 2$ and $\alpha = 2$.

Beckers et al. (1992b) offered *Lasius niger* ants two identical bridges between food and nest, after 30 minutes, the majority of the ants took only one of the two bridges (figure 3.1d). Sumpter & Beekman (2003) reported similar results for Pharaoh's ants when they offered the ants two identical feeders in opposite directions from the nest. Instead of a 50:50 split between feeders, the split was closer to 70:30 or 30:70, giving a u-shaped distribution of number of ants at one of the feeders over all the trials. The "winning" feeder was the one that had the most ants nearby when it was initially placed in the foraging arena (Sumpter & Beekman 2003).

The emergence of an asymmetrical distribution of individuals in a uniform environment is a characteristic property of positive feedback (Camazine et al. 2001; Deneubourg & Goss 1989; Pasteels et al. 1987). Whether a recruitment system will exhibit such symmetry breaking depends on the strength of the positive feedback. Figure 3.3b shows that when the total number of foraging ants is low the model predicts an even split of ants between food sources. When the total number of foraging ants reaches a critical value this symmetry is broken and one food source is chosen almost exclusively. These symmetry breaking bifurcations arise in many situations where the response is disproportional to the difference between two signals (i.e., $\alpha > 1$).

A symmetry breaking bifurcation is also predicted to occur as the amount of pheromone laid per individual foraging ant is increased. This prediction was confirmed by Portha et al. (2002). They found that *Lasius niger* engage in a higher intensity of pheromone laying when feeding on a sucrose food source than on a protein food source. As a result colonies fed on carbohydrate showed a stronger tendency to break the symmetry between two feeders than those fed on protein.

Portha et al.'s (2002) results allow us to explain symmetry breaking in terms of the ants' need to respond differently to different types of food. Under natural conditions, *Lasius niger* collect sucrose by extracting honeydew from a limited number of aphid colonies. These are long lasting food sources that require defense against competitors and predators, and are thus best exploited by a concentrated response by large numbers of foraging ants. Conversely, natural protein sources take the form of dead insects that are not usually spatially clustered and require only small groups of ants to retrieve. By adjusting their individual pheromone laying behavior in response to food type the ants regulate their collective foraging response to that required to deal with that particular food. These observations provide a functional explanation for what may at first appear as a mathematical oddity of the model in box 3.A. Symmetry breaking occurs when it is beneficial for the majority of individuals to make the same choice.

The model presented in box 3.A makes a number of simplifications, which mean that while improving our qualitative understanding of patterns, it does not give accurate quantitative predictions about distribution of foragers between routes or food sites. More detailed and quantitatively accurate models have been developed of the foraging of Argentine ants (Goss et al. 1989) and *Lasius niger* (Beckers et al. 1992a; Beckers et al. 1992b; Beckers et al. 1993). In the latter case, ants taking the longer bridge were found to be more likely to make U-turns than those following the shorter bridge (Beckers et al. 1992b). When taking this additional bias into account, reasonably accurate quantitative estimates of the distribution of ants between bridges can be made (Camazine et al. 2001).

Honeybee Dances

Probably the most celebrated mechanism for transferring information about food is the waggle dance of the honeybee (Seeley 1995; von Frisch 1967). Waggle dances are performed by honeybee foragers that have successfully found nectar or pollen. The dance is a figure eight pattern: a waggle run, where the bee vibrates its abdomen and wings as it walks forward, followed by a turn to the right circling back to the point at which the run begun, followed by another waggle run and a further turn back, this time to the left. The direction and the duration of the run are correlated with the direction and the distance from the bee hive to the food source. Uninformed bees in the hive follow a dance and then fly in the direction of and for the distance encoded by the dance, after which they search locally using odor and visual cues (Riley et al. 2005). Usually this recruited bee will fail to find the advertised food site, but by repeatedly returning to the dance floor and following further dances she will eventually find and return with food (Seeley & Towne 1992). Since recruited bees may later perform dances themselves the waggle dance acts as a positive feedback mechanism through which information about food is transferred.

We can put the waggle dance in the framework of our model in box 3.A. If only two food sources equidistant to the hive, X and Y, are available to the colony then the probability that an unemployed bee is recruited to feeder X can be expressed as

$$\frac{x+k}{x+y+2k}, \tag{3.2}$$

where x is the number of bees dancing for site X, y is the number of bees dancing for site Y, and k is a constant (Sumpter & Pratt 2003). This equation reflects the observation that choice of foraging site is directly proportional to the level of dancing for that site (Seeley & Towne 1992). When the total number of dancing bees, $x+y$, is small then the probability of going to either site is close to one-half, consistent with the observation that honeybees search independently for food in the absence of dance information (Beekman et al. 2007; Seeley 1983).

Equation 3.2 is a specific example of equation 3.1 with $\alpha = 1$. Honeybees are known to retire from foraging at a rate inversely proportional to a feeder's profitability, so the parameters q_x and q_y in box 3.A can be set to reflect the relative profitability of feeders X and Y. Figure 3.4a shows a bifurcation diagram for the case where two feeders have unequal profitability. The majority of bees forage at the most profitable site. Unlike the comparable bifurcation diagram in figure 3.2c there is, for any flow of foragers out of the hive, a unique steady state number of foragers going to each site. Further analysis in box 3.A shows that, independent of the quality and number of food sources, there is a unique steady state for all initial conditions. Similarly, figure 3.4b shows that when both feeders are of equal quality honeybees never exhibit symmetry breaking. The bees will always divide their workforce equally between two equally profitable feeders.

Although not extensively tested for different foraging scenarios, the above model's predictions do seem to hold in experiments. Seeley et al. (1991) offered honeybees two feeders of different quality in opposite directions from the hive and after four hours swapped the feeder quality. The bees were able to track this change and re-allocate their workforce appropriately. Although the subject of theoretical debate (Camazine & Sneyd 1991; de Vries & Biesmeijer 2002), symmetry breaking for two equal feeders has never been tested directly for honeybees. Bartholdi et al. (1993) has, however, shown that the ratio of recruitment over retirement equilibrates when a colony is offered two feeders of different quality, i.e., differing concentration of sugar solution, but with limited capacity, i.e., the number of bees per minute that can access the feeder. The overall picture of honeybee foraging is one of decentralized tracking of the environment, with each honeybee a relatively complex information storage unit that uses dance language to share details of how the environment changes (Biesmeijer & de Vries 2001; Seeley 1997; Seeley 2002).

This picture of honeybee foraging contrasts with the picture of foraging of *Lasius niger* and other ants, where large numbers of ants are quickly mobilized to lock into a particular sucrose food source. We have already seen that *Lasius niger* colonies adjust the degree of symmetry breaking in their foraging to match the properties of the food they are collecting. In this context, it is important to note that by sampling more than one dance and taking directional information from the dance for which the majority of sampled bees are dancing, an uninformed forager could bias its probability to forage at the most popular site to be greater than the proportion of dances for that site. Such sampling would then produce a symmetry breaking in the distribution of foragers between sites (Camazine & Sneyd 1991). Whether such repeated sampling has not evolved

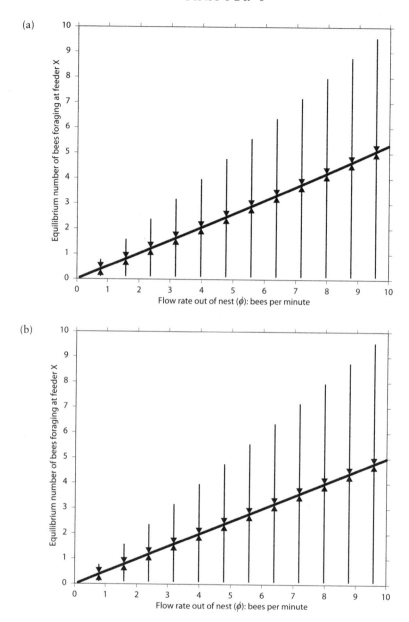

Figure 3.4. Bifurcation diagrams for honeybee foraging model ($\alpha = 1$ in the model in box 3.A), (a) when feeders have unequal profitability, $q_x = 1$ and $q_y = 0.95$; and (b) when feeders have equal profitability, $q_x = q_y = 1$. There is a small difference in the position of the equilibrium that is difficult to see. It is interesting to contrast this small difference with the large difference in the bifurcation diagrams in figure 3.2c and figure 3.3b. Other parameter values are $v = 1$ and $k = 2$.

because of physiological limitations at the level of individual foragers or because the environment honeybees typically experience is not usually highly clustered is an interesting, but difficult to answer, question.

As with ant foraging, more detailed models have been built of honeybee foraging. Based on and parameterized by Seeley's experiments, Camazine & Sneyd (1991) developed the first differential equation model of honeybee foraging. This was developed further and put in the general context of social insect foraging by Sumpter & Pratt (2003). De Vries & Biesmeijer (1998) developed an individual-based model of honeybee foraging and pointed out a number of limitations in the differential equation models in making accurate quantitative predictions about foraging. De Vries & Biesmeijer (2002) suggest that symmetry breaking does occur in their individual-based model. However, rather than finding two symmetrical stable distributions of foragers between food sources, the maximum asymmetry (i.e., the maximum absolute difference between the numbers of foragers visiting the two sources) differed greatly between simulation runs with the same parameter values. A similar dependence on initial conditions is seen in the differential equation model of honeybee foraging when $k = 0$ and is due to the maintenance, rather than the amplification, of a fluctuation in the early between-feeder distribution of foragers. I would thus challenge the assertion that the individual-based models demonstrate symmetry breaking. Indeed, the detailed individual-based model loses out to the simple model in box 3.A in terms of generality, while providing little additional understanding of positive feedback. Individual-based models can, however, prove more powerful in understanding the multiple feedbacks inherent in complicated co-operative systems, and I will return to such models in chapter 9.

Other Signal-based Recruitment

Recruitment signals made at or emanating from a central nest are not limited to honeybees or pheromone trail laying ants. Stingless bees show an array of contact-based, visual, scent-based, and acoustic communication signals that allow foragers that have found food to recruit those that have not (Biesmeijer & Slaa 2004; Nieh 2004). Many species of ants exhibit group and tandem-running recruitment where, after a signal made in the nest to attract would-be followers, an ant that has found food leads recruits directly there (Franks & Richardson 2006; Hölldobler & Wilson 1990). Symmetry breaking is observed in tent caterpillars. When offered a choice between two equal food sources, a caterpillar colony will aggregate on only one of the two (Dussutour et al. 2008; Dussutour et al. 2007).

Recruitment signals are not limited to insects. For example, Norway rats deposit odor trails from the food back to the nest (Galef & Buckley

1996). By attracting nestmates, these trails spread information about widely scattered, ephemeral food sources, reducing the time it takes individuals to find food (Galef & White 1997). Naked mole rats also leave odor trails on finding food, make chirping noises during their return trip, and display the collected food for nestmates (Judd & Sherman 1996). There is evidence for a weak form of positive feedback with follower naked mole rats vocalizing when they find food, but with a lower probability than the initial discoverer. Recruited mole rats appear to look for the trail left by a specific individual, suggesting that recruitment to a particular food source is proportional to the number of recruiting individuals. However, chemical signals and food calling both have the potential to generate disproportional recruitment and symmetry breaking.

Cue-based Recruitment

In all of the above examples, information exchange about the location of food has involved a signal from an informed individual to an uninformed individual. A signal is defined as "an act or structure that alters the behavior of another organism, which evolved because of that effect, and which is effective because the receiver's response has also evolved" (Maynard Smith & Harper 2005). Information can also be exchanged by cues, which are "a feature of the world that can be used by the receiver as a guide to future action" (ibid). Such features might be an aspect of an informed individual's behavior, such as the direction it moves, but can also be a way in which the individual has modified its environment, such as leaving footprints or carrying an odor. Rather than evolving to communicate the existence of a resource, cues arise when a particular behavior happens to be correlated with obtaining a particular resource. Uninformed individuals then use this correlation to gain information. The cue may incur a cost to the informed individual in terms of increased competition for the resource, but not as a result of the behavior itself.

There are a number of examples of information transfer by cues. Ratcliffe & Hofstedeter (2005) showed that observer bats that first interacted in the nest with a "demonstrator" bat chose the same food type as that eaten by the demonstrator. Since the observers experienced novel food cues only on the breath or body of the demonstrator, the experiments suggest that the interactions served to induce the bats to search for a particular type of food source. In this case the information exchanged was simply about the existence of a particular food rather than its location. By following informed nestmates, however, an uninformed individual can ascertain the location of food without the need for signals. Such following of informed individuals is seen in hooded crows (Sonerud et al. 2001) and cliff swallows (Brown 1986).

A problem with inferring that information exchange is solely cue- rather than signal-based is that it is difficult to completely rule out the existence of a signal. For example, ravens certainly use cues, but possibly also use signals, in information transfer at their communal roosts. When naïve North American ravens were added to communal roosts they fol- lowed their informed roost-mates to new feeding sites (Marzluff et al. 1996). At the beginning of these flights some birds produce noisy "kaws" and "honks," although it is not known whether these noises are more often produced by informed birds. There is, however, evidence based on a small number of observations of European ravens that the first birds to be seen at a bait carcass were also those that performed flight displays and vocalizations the evening before and appeared to initiate morning departures from the roost (Wright et al. 2003). These observations would suggest that informed ravens actively signal the location of food. Other animals may use only cues at the nest, but use evolved signals once they arrive near food. For example, although not known to use any signals to recruit from the nest, cliff swallows are known to use a vocal signal when they arrive at food that alerts other nearby swallows of its location (Brown et al. 1991).

Foraging Success and Group Size

When food is difficult to find then an individual using information pro- vided by others can increase its rate of finding food. The honeybee dance improves efficiency of food collection during seasons and in environ- ments where forage is clustered (Dornhaus & Chittka 2004; Dornhaus et al. 2006; Sherman 2002). Likewise, Brown et al. (1991) suggest that cliff swallow signals about food location occur only when the insects upon which they feed are spatially clustered.

Information transfer can produce a synergism for group members (see Synergism in chapter 10). Specifically, it can lead to increasing per capita foraging success with group size. Brown & Brown (1996) provide evi- dence for this in cliff swallows, where both the total number of food de- liveries to brood per parent per hour and the amount of food collected per foraging trip per parent increases with group size (figure 3.5a). Stud- ies of the Pharaoh's ant (Beekman et al. 2001) and Argentine ants (Hal- ley & Burd 2004) also show that the per capita number of ants arriving at a feeder increases with colony size (figure 3.5b). In general, however, the study of per capita productivity in insect societies has been mainly focused on the early stages of colony foundation, where increases in pro- ductivity are usually attributed to co-operative building (see Tunnels and Tents in chapter 7) and defense than to foraging success (Bernasconi et

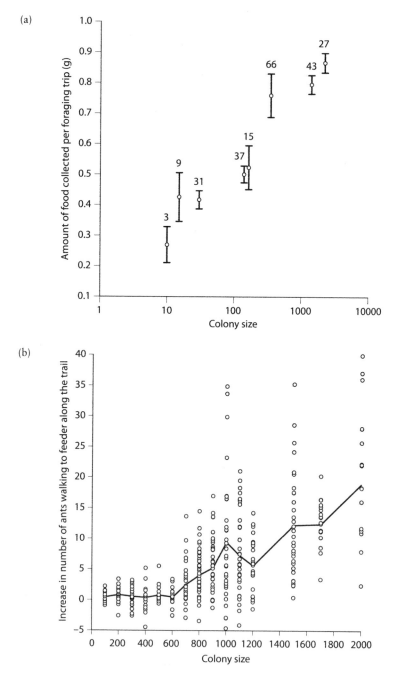

Figure 3.5. How foraging success increases with group size for (a) cliff swallows (adapted from Brown & Brown 1996) and (b) pharaoh's ants (adapted from Beekman et al. 2001).

al. 2000). One study of the early stages of colony foundation that could relate to information transfer looked at brood raids by fire ants on other nearby ant colonies. Adams & Tschinkel (1995) found that nests consisting of multiple queens produced more workers and then had an increased success during raids on other colonies.

While information transfer can lead to per capita gains in foraging success, it can require a minimum number of individuals to function effectively. If an ant in a small colony finds a food source a long way from the nest, then by the time another ant passes over the place she left pheromone trail, the pheromone will probably have evaporated. In this case, the trail does not help other ants find the food. For large colonies of ants, however, it is more likely that an ant will find the pheromone trail before it evaporates, follow it, and thus reinforce it.

Beekman et al. (2001) formalized this argument in a differential equation model, similar to that in box 3.A, of trail laying to a feeder. The model predicted that (a) as the number of ants in the colony increased the number of ants visiting the feeder would increase non-linearly and (b) provided the rate at which ants found the food without following a trail was small, then at a critical colony size there would be a sudden switch from few ants visiting the feeder to a large proportion of the ants visiting the feeder. This prediction was confirmed experimentally for ants foraging at a single feeder (Beekman et al. 2001). Small colonies of *Monomorium Pharonis* were unable to establish an effective pheromone trial, while above a critical size trails were formed between nest and food (figure 3.5b). These observations could help explain why pheromone trail laying has evolved primarily in ant species that contain large numbers of workers (Beckers et al. 1989) and that ants of the same species change their trail laying behavior dependent on their colony size (Devigne & Detrain 2002).

Evolution of Information Centers

Synergism and Altruism

The reason ᶦᵗ is important to draw a distinction between cues and signals in information transfer is that signals incur an efficiency cost to the informed individual producing them (Guilford & Dawkins 1991). The cost can be either a direct result of the time or energy expended in making the signal, e.g., in performing a dance or producing pheromone chemicals, or a result of increased competition for the signaled resource. In order for a costly signal to have evolved there must also be an associated benefit. This benefit must on average be greater than the cost. The key evolutionary question about all systems where we see signaling is: what are the benefits of signaling?

Such questions do not usually have one simple answer but depend on a whole range of factors. In chapter 10, I discuss three general settings under which co-operative signals can evolve in spite of the possibility that other individuals could cheat by following others' signals while not producing their own. These are repeated interactions, synergism, and inclusive fitness.

Although information centers involve repeated interactions, a central requirement for this type of co-operation—that individuals are able to identify one another—is not usually fulfilled (Repeated Interactions in chapter 10). Instead, one or more of synergism and inclusive fitness are likely to be the most important factors in the evolution of information centers. The key idea in synergism is that although individuals pay a cost in signaling the location of food, the fact that all individuals in the group produce this signal provides a per capita benefit that outweighs the cost. In particular, provided per capita foraging success increases at least linearly with group size, synergies can evolve even if it would not pay an individual to start signaling in a group of non-signalers (Synergism in chapter 10). The key idea of the inclusive fitness argument for co-operative signaling is that signals that increase the chance of genetically related individuals finding food provide indirect fitness benefit to the focal individual (Inclusive Fitness in chapter 10).

Cliff swallows do not nest in colonies of related individuals and inclusive fitness plays little or no role in the evolution of their foraging behavior (Brown & Brown 1996). Correspondingly, there is a lower degree of signaling between colony members than for social insects and communication about food location is primarily cue-based. However, signaling between birds is seen in the form of food calling at mobile insect swarms. The signaling birds can track the swarm while being able to make return journeys between the colony and the insects. A functional explanation of cliff swallow foraging based on synergism is supported by the per capita foraging success of these birds, which increases with colony size (figure 3.5a).

Many social insect species have a high degree of within-colony relatedness, and there is little doubt that inclusive fitness contributes to the co-operation inherent in these species (Bourke & Franks 1995; Foster et al. 2006). Several authors have argued that because relatedness within these colonies is lower than first predicted, inclusive fitness may have a less important role in co-operation than once supposed (Korb & Heinze 2004; Wilson & Holldobler 2005). It is here that the observation that signaling in foraging increases per capita foraging success with group size plays an important role. Synergism leads to an increase in benefits and thus a lower requirement for within-colony relatedness for the evolution of co-operation. Similarly, for naked mole rats, the relevance of high within-group relatedness (Reeve et al. 1990) has been questioned because the degree of competition between relatives has not been measured (Griffin & West 2002). With or without competition for resources, synergism whereby co-operation increases the amount of available resources could lead to the evolution of signaling during foraging. Further empirical tests of the foraging performance of different sized colonies are needed to clarify the role of synergism in these species.

The system that is possibly most difficult to provide a functional explanation for co-operative signaling are the flight displays by ravens (Wright et al. 2003). For these birds, groups may be sufficiently small that repeated interactions, either in terms of direct reciprocation or indirect reputation building, could play a role (Repeated Interactions, chapter 10). However, these groups are relatively fluid and it would be interesting to have more data on the probability of repeated interaction needed to justify reciprocation.

Social Parasitism

Cues can be thought of as unavoidable consequences of possessing information. For example, bats that carry the smell of food also carry information about its existence. Similarly, it is difficult for a bird to fly to food without revealing to others where it is going. In these cases the informed individual may pay some cost associated with being followed, in terms of increased competition for food. However, because the cue is not an evolved communication mechanism, we no longer need to find an associated benefit with information transfer. Instead, cues are an example of social parasitism (Parasitism, chapter 10). The bat who collects food gathers information and those back at the nest parasitize that information.

One question that now arises is whether the informed individual might evolve some mechanism to disguise the information it possesses, and thus avoid paying competition costs. Disguising of information has not been observed in the systems discussed in this chapter. For example, cliff swallows show no signs of disguising the fact they have found food (Brown 1986) and away from the nest they actively recruit other individuals to food. Given the possible benefits, or at most small costs, of information sharing for cliff swallows, the lack of disguising behavior is hardly surprising. Indeed, the systems I have described are chosen precisely because they showed some form of information transfer and are not likely candidates for observing hiding behavior. Strategies to disguise the position of food do, however, occur in other species, such as hording by scrub jays (Emery & Clayton 2001), where there are large costs to sharing information.

Exploiting the Finds of Others

Information transfer does not always occur at or originate from a central point. There are many situations where animals copy the choices of others that have information about food, mates, or shelter (Dall et al. 2005; Danchin et al. 2004; Wagner & Danchin 2003). In some cases, observers can gain information directly about the quality of the environment from the success or otherwise of others. For example, starlings use observations of their flockmates' success in probing for food to decide when to leave a patch in search of another (Templeton & Giraldeau 1996; Templeton & Giraldeau 1995). In other cases, copying can occur without an obvious way in which the observer can assess the success of the participant. For example, female quails show a tendency to mate with males that they have previously seen mating with other females (White & Galef 1999a; White & Galef 1999b) and even prefer other males that share the characteristics of a male they have seen mating (White & Galef 2000).

Simply copying the behavior or blindly following others can allow animals to make better choices, even if they are unable to assess the quality of the information possessed by the copied individual. For example, consider the options available to a foraging bird when it arrives at a field and sees one other bird with its head down in the ground pecking for food. From this observation alone the observer is unable to assess with certainty whether the pecking bird has found food or is simply looking. However, the pecking bird is more likely to have its head down if it has found something. Thus the proportion of time the bird has its head down is a good indicator of the pecking bird's success. The observer need not know exactly how long a bird has had its head down, but can gain an instantaneous estimate of its foraging success simply from whether its head is down or not at the time of arrival. A simple rule for deciding whether to join could be as follows: always join a bird with its head down, never join a bird with its head up. If the observer applies this rule whenever it arrives in a field it will, on average, do better than if it joined or not at random. Experiments where geese flocks were presented with models of an artificial flock of model birds, some with heads down and others with heads up, show that flocks are more likely to land near groups of models where more birds have their heads down (Drent & Swierstra 1977).

There is evidence for increased foraging success through group membership. Grunbaum & Veit (2003) found that albatrosses spent a larger proportion of their time feeding when in larger groups. With these observations it is difficult to separate cause and effect. Is the larger group simply due to more abundant food at certain points in the environment? However, Grunbaum and Veit found only a weak relationship between density of albatrosses and the density of available krill, suggesting that foraging success had a stronger positive relationship with the number of birds foraging than with prey density. More direct evidence is available from experiments on fish. In experiments where food was available in only one of a large number of pots, the time it took goldfish and minnows to find food decreased roughly in proportion to one over the group size (Pitcher et al. 1982). This would suggest that the transfer of information is highly effective in these groups, with finds rapidly communicated between individuals.

Producers and Scroungers

Before we can sensibly discuss information transfer by animals foraging in loosely formed groups we have to ask when it is beneficial to copy the behavior of others. When food items are highly clustered, relying solely on your own independent search is not always the best strategy. If another individual finds a cluster of food, then it pays to join that individual and take a share in the find rather than continuing an independent search. On the other hand, it is usually difficult to watch others and search at the same time. If everyone spends all their time watching what others have found then no one will ever find anything. This dilemma has been posed as the producer-scrounger game.

The producer-scrounger game and other related models (box 3.B) predict that—assuming (1) food patches are large enough so that they cannot be quickly consumed by one individual and (2) there is some cost to copying others in terms of lost possibilities of independently finding food—some proportion of a group will join others that have found food rather than searching themselves. The models further predict that when food patches are larger, or food is more patchily distributed, the proportion of observations of joining behavior will increase. These predictions have been shown to hold for spice finches (Giraldeau & Beauchamp 1999; Giraldeau & Livoreil 1998), where joining behavior increases with the patchiness of food distribution (figure 3.6).

The producer-scrounger game can be further interpreted as predicting that group members will learn to adopt one of two alternative strategies: producing, where individuals look for food independently; or scrounging, where individuals copy the finds of others. While changes in finding and joining rates with food distribution provide evidence that the birds have some mechanism to tune their social behavior to better exploit their environment, it does not in itself constitute evidence that birds have learned to adopt these alternative strategies. To test whether spice finches learn to adopt appropriate strategies, Mottley & Giraldeau (2000) set up a barrier, one side of which allowed "producers" to open a foraging

Box 3.B The Producer-Scrounger Game

In the basic producer-scrounger game individuals can choose to adopt either one of two distinct strategies: *producers* search the environment and find food clumps at a constant rate (Giraldeau 2000; Giraldeau & Beauchamp 1999; Ranta et al. 1996; Vickery et al. 1991). They get a finder's share α of the food they find, but are unable to consume a further $(1-\alpha)$ before the *scroungers*, which do not search themselves but instead watch the producers, arrive at the find and divide the remaining share between themselves and the finder. In general, $1-\alpha$ will increase with the size of a food patch. The average rate of food intake or, in game theory terms, the payoff of the producer in a group of size N of which a proportion s are scroungers is

$$w_P(s) = \alpha + \frac{(1-\alpha)}{(1+sN)}.$$

The payoff for scroungers is

$$w_S(s) = \frac{(1-\alpha)(1-s)N}{(1+sN)}.$$

From these payoffs we can see that when the population consists purely of producers, i.e., $s = 0$, then $w_P(0) = 1$ but $w_S(0) = (1-\alpha)N$, so provided $1-\alpha > 1/N$, it always pays to be a scrounger. Likewise, in a population purely of scroungers $w_S(1) = 0$ while $w_P(1) > 0$, so it always pays to become a producer. By solving $w_P(s_*) = w_S(s_*)$ for s_* we can find the evolutionarily stable proportion of scroungers in the population (see chapter 10 or Giraldeau (2000) for details). This is $s_* = (1-\alpha) - 1/N$. The group will consist of some proportion of scroungers and some producers, the proportion of scroungers increasing with the size of the available food patches.

Many bird species have a visual field that allows them to simultaneously scan for scrounging opportunities while searching the ground for food (Fernandez-Juricic et al. 2004a). This observation can be accounted for as the opportunistic strategy, where we assume $\beta < 1$ is the rate at which opportunists find food while still able to scrounge whenever an opportunity arises (Vickery et al. 1991). This strategy has payoff

$$w_O(s) = \beta\alpha + \frac{\beta(1-\alpha)}{(1+sN)} + \frac{(1-\alpha)(1-s)N}{(1+sN)} = \beta\alpha + \frac{(1-\alpha)(\beta+(1-s)N)}{(1+sN)}$$

where s is now the proportion of opportunists. The evolutionarily stable proportion of opportunists is $s_* = (1-\alpha)/(1-\beta) - 1/N$. Thus, opportunists will make up a larger proportion of the population than scroungers. However, provided there is a cost to being an opportunist (i.e., $\beta < 1$) then the producer strategy can always invade a group of pure opportunists. Similar predictions hold for other game theory models of how individuals in groups can parasitize information about the location of food (Clark & Mangel 1984; Ranta et al. 1996; Ruxton et al. 2005). Provided there is a cost to opportunism or scrounging then this strategy will coexist with that of producing, and joining behavior will increase with the size of food patches. In general, I would classify all of these games as capturing a form of social parasitism (chapter 10).

In the above discussion I have used the producer-scrounger model to make functional predictions about how foraging strategy will change with food patch size. It is possible, however, to further interpret the producer-scrounger model as a mechanistic model. In this case, it predicts that individuals in a group will switch between two distinct types of behavioral strategies until they reach the evolutionarily stable strategy (Giraldeau & Beauchamp 1999). Much of the experimental research on producer-scrounger games has concentrated on this interpretation and has identified situations where some individuals adopt a scrounging strategy and some individuals a producing strategy. However, this mechanistic interpretation is not necessary for many of the key predictions of producer-scrounger models to hold.

compartment that was then also accessible to "scroungers" on the other side. They found that the proportion of finches on the scrounging side quickly converged to that predicted by the stable state of the producer-scrounger model. However, the setup for these experiments was somewhat artificial, with the barrier forcing the birds to choose one of two distinct strategies.

Establishing that distinct strategies occur in natural foraging environments requires identifying specific behavioral features to be associated with particular strategies. Barnard & Sibly (1981) found evidence for

this in an experiment on house sparrows. They noted that birds that found more food through independent searching performed "a characteristic zig-zag hopping and . . . frequent head-cocking," while those that found more food through interactions with others either remained still or hopped directly towards other birds. Coolen et al. (2001) found similar correlations for spice finches. Hopping with the head pointing down was correlated with finding, while hopping with the head up was correlated with joining. There was also evidence that the finches learned to adapt their strategy to their environment. When offered a seed distribution where patches contained only one seed, i.e., in which scrounging was a poor strategy, joining was no longer observed after four days and after six days hopping with the head up also decreased to zero.

Quorum Mechanisms for Information Transfer

While birds of some species may adopt alternative producer-scrounger strategies, the existence of distinct strategies is not a requirement for information transfer. One simple strategy that allows individuals to exploit food finds of others is copying. Behavioral responses whereby an animal's probability of exhibiting a particular behavior is an increasing function of the number of conspecifics already performing this behavior, are a common feature of animals that form groups (Sumpter 2006). Collins & Sumpter (2007) looked at the feeding patterns of commercially farmed chickens. We found that the probability that a bird starts feeding at a particular point along a feeding trough was an increasing function of the number of birds already feeding there (figure 3.7a) and that the probability that a bird stops feeding at a particular point was a decreasing function (figure 3.7b).

Copying often takes the form of a *quorum response*, where the probability of performing an action sharply increases when a particular group size, or quorum, is reached (Sumpter & Pratt 2008). Box 3.C gives a simple example of a quorum response model of bird feeding dynamics. The probability of taking a particular action is a sharply increasing non-linear function of the number already performing it and repeated interactions lead to positive feedback. This model exhibits both the symmetry-breaking and the potential for enhancement of sub-optimal choices seen in the ant foraging model (box 3.A). This model is studied in more detail in chapter 6. Here, I compare the predictions of quorum-response models to data on bird foraging.

The observation in figure 3.7 of how rates of finding and joining change with food distribution can be explained using the quorum response model. The model in box 3.C predicts an increase in joining behavior with food

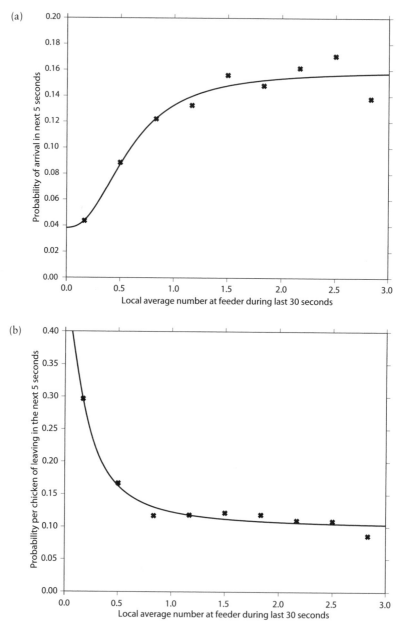

Figure 3.7. How the rates of arrival and leaving of chickens change as a function of the number of other chickens already at a particular point along a feeding trough. Measured frequencies of (a) arrival and (b) leaving the feeder as a function of a moving average of the local density at a section on the feeder. The solid lines are fitted response functions. See Collins & Sumpter (2007) for details.

Box 3.C Quorum Response Model of Bird Feeding

The model of Collins & Sumpter (2007) uses the rates of joining and leaving (shown in figure 3.7) along with those for moving along the feeder measured from observations of chicken feeding, to predict the dynamics of feeding over time. Here I describe a simpler version of that model in which only the probability of joining depends on the number of individuals at a food patch and there is no explicit spatial structure to the patches. In this model I assume that there are f distinct food patches and n birds. Let $C(i, t)$ be the number of birds at patch i at time t and $B(t)$ be the number of birds that are not at a food patch. Initially, $B(0) = n$ and $C(i, 0) = 0$ for all i. On each time step, the probability per bird not at the food patch of arriving at food patch i is

$$s + (m - s)\frac{C(i,t)^\alpha}{k^\alpha + C(i,t)^\alpha} \tag{3.B.1}$$

where s, m, α, and k are the constants: s is the probability per time step that a bird arrives at the patch in the absence of other birds, m is the maximum probability per time step that a bird arrives at the patch, k is the threshold number of birds at which the probability of arrival at the patch is $(s + m)/2$ and α is the steepness of this threshold. The probability per bird at a food patch of leaving is constant l.

To model natural foraging conditions, where food is limited, we set a constant probability, p, per time step that d food units appear at a patch. The birds arriving at the patch eat one unit of food per time step. Once all the food at that patch is eaten all the birds leave. In order to simulate different levels of food clustering we can change d. The larger the value of d, the more clustered the distribution of food. Setting $p \propto 1/d$ ensures that on average a constant amount of food is available. Figure 3.8a shows how the rate of food intake changes, and figure 3.8b shows how the rate of finding and joining changes with the degree of food clustering. In the model, and consistent with experimental observations of spice finches (Giraldeau & Livoreil 1996; Giraldeau & Beauchamp 1999), as clustering of food increases, so too does the frequency of joining behavior.

patch size. The key idea here is that when food patches are larger, the birds that find them stay there longer. If other birds have a probability of joining that increases with the number of other birds at a patch, then larger food patches will attract more joiners, leading to a positive feedback loop whereby joining becomes still more common. The result is that the individuals using a quorum response will increase their joining rate with the degree to which food is clustered (figure 3.8). Instead of requiring the birds to learn a strategy in response to food distribution, the quorum response automatically tunes joining rate to patch distribution.

The quorum response model is a mechanistic explanation, while the producer-scrounger model is primarily a functional explanation. When these two approaches are combined they give a very powerful framework for thinking about social parasitism in foraging groups. The producer-scrounger model identifies a stable strategy for individuals foraging in a group, and the quorum response model shows how that equilibrium can be reached by individuals that do not have complete information about the structure of the environment in which they live. The quorum response model gives a simple but plausible explanation of how individuals can tune their response to optimize the intake of food in variable environments. It suggests that animals can achieve this balance without having to learn to adopt particular strategies based on previous experience.

It would be nice to be able to interpret the chicken data in terms of information transfer among animals that have evolved in natural environments. However, farmed chickens have been bred for rapid growth over many generations (Weeks et al. 2000). This breeding may have led to increased copying behavior, which in turn would lead to increased feeding and thus more rapid growth. Thus copying in chickens may tell us more about the qualities farmers want to see in chickens and less about how they use the behavior of others to gain information about their environment.

Further studies are needed to quantify copying responses in the foraging of other birds than farmed chickens. Such studies could produce interesting results in a research area where it is difficult to conduct clear-cut experiments (Giraldeau & Beauchamp 1999). In parallel with experiments, a more thorough combination of producer-scrounger theory with mechanistic descriptions of how individuals react to the behavior of those nearby would shed light on the question of when and where social parasitism will be observed. The quorum response model has strong similarities with models of social insect foraging, described in box 3.A. The self-organized tuning of response to patch size is similar to that achieved by ants and honeybees when tuning their collective response to food quality. The difference here is that the individual, not the colony, should be thought of as the optimizing agent. Thus, while it is interesting that a simple mechanistic

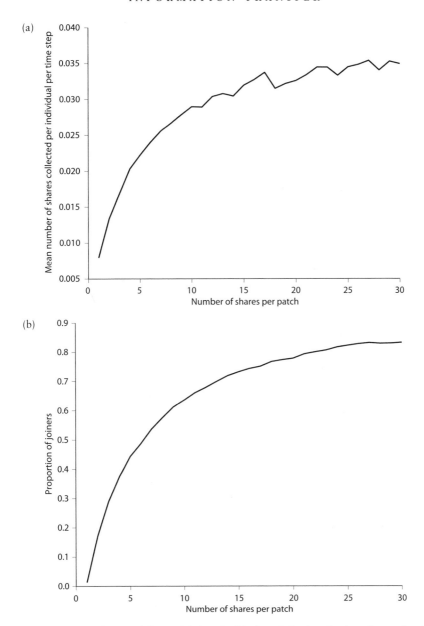

Figure 3.8. Simulation of the model described in box 3.C when food at the patches is limited. (a) Intake per forager and (b) proportion of foragers finding and joining changes with the degree of spatial clustering of food. Here, $n = 25$, $f = 10$, and $p = 0.1/d$. As the number of food items appearing each time food becomes available, d, increases, so too does the degree of clustering of the food. Other model parameter values are $s = 0.001$, $m = 0.4$, $k = 1$, and $\alpha = 4$.

model appears to reach the stable equilibrium of the producer-scrounger model, further theoretical investigation is needed to test whether particular response thresholds could be exploited by other alternative strategies.

From Copying to Culture

A good theory of the reasons for and the effects of copying could have consequences that go beyond understanding chickens scrabbling in farm-yard dirt. It could reveal something about the fads, the fashions, and even the religions of our own society. Conventions and customs of our society are established by information transfer among individuals and it is plausible that the roots of these social norms lie in something as simple as copying the behavior of others.

A classic experimental demonstration of copying behavior by humans was conducted by Milgram et al. (1969). On a busy New York street they placed a small stimulus crowd of individuals, each of which looked up at a window of a nearby building. They then observed passersby as they walked past the crowd. They found that the larger the crowd the larger the proportion of passersby who would stop and/or look up. Hale (2008) repeated these experiments in Oxford and found similar results, although with a weaker response by passersby. Figure 3.9a, b shows how proportion of passersby looking up increases with crowd size in both cases. In both cases the functions relating proportion looking up to crowd size are initially approximately linear (i.e., $\alpha \approx 1$ in equation 3.1), and not the quorum-like (i.e., $\alpha > 1$) relationship in, for example, the ants' response to pheromone.

Quorum-like responses are also observed in humans. Another classic experiment by Asch (1955) looked at individuals in situations where they felt under social pressure to conform. He showed subjects two cards. One card showed a single line of a "standard" length and the other showed three different length lines, only one of which was the standard length. He asked the subjects to identify the standard length line. When on their own, individuals nearly always successfully identified the standard line. When a single subject was placed in a group of "opponents," who were instructed beforehand to deliberately choose the same but incorrect line, the subject would often concur with the opponents. Figure 3.9c shows how the proportion of individuals making an error increases with the size of the group of opponents. In this case the response is a sharp quorum threshold at approximately 2 individuals. Having two or more opponents leads individuals to make mistakes.

Both linear and quorum-like responses have been the focus of a great deal of theoretical interest in sociology and economics. Granovetter

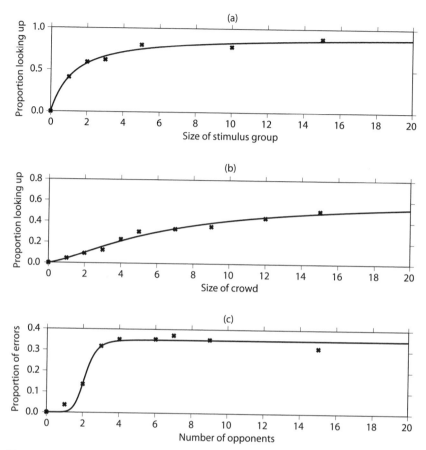

Figure 3.9. Responses of humans to the behavior of others. The relationship between the probability that passers-by will copy the gaze of the stimulus group as a function of stimulus group size, in (a) New York (Milgram 1968) and (b) Oxford (Hale 2008). (c) The probability that an individual will concur with a group of opponents as a function of the number of opponents (Asch 1955). The fitted line is the function $P(N) = m(N^k/T^k - N^k)$, where $P(N)$ is the observed frequency of looking up and N is the group size (opponents or crowd). The fitted parameters, T, m, and k characterize the type of response: m is the maximum proportion of individuals that will look up, T is the threshold group size at which $m/2$ individuals will look up, and k determines the shape of the functional response. The parameters are (a) $m = 0.63$, $T = 6.4$, and $k = 1.42$; (b) $m = 0.91$, $T = 1.2$, and $k = 1.05$; and (c) $m = 0.35$, $T = 2.13$, and $k = 6.66$.

(1978) describes a model where individuals decide whether to engage in some form of action (such as a rioting, use of contraception, or voting for a particular party) when a quorum threshold of others have already engaged in the action. He showed that groups with similar average preferences may generate very different collective behavior, depending upon the order in which they make the decision. Schelling (1978) proposes similar models and introduces concepts like "tipping points" and "critical mass" to describe how social activities suddenly take off when, through some essentially random fluctuation, a threshold is passed.

At some point copying can become culture. Ball (2004) provides an excellent summary of how the "critical mass" idea has been applied in the study of crime, wars, and economics. For example, sudden increases or drops in crime may be attributable to the passing of a critical threshold at which criminal behavior is socially "acceptable" (Ormerod et al. 2001). Similarly, Skog (1986) studied long term changes in consumption of alcohol in Norway and showed that it was consistent with a simulation model of changes due to social interactions. Saam & Sumpter (2008) showed that these models could explain decision-making by nation states during treaty negotiations.

It is very hard to disentangle cause and effect when dealing with sociological phenomena. There are lots of rapidly changing external factors, such as economics, population levels, and large scale social change, which correlate with crime levels and alcohol consumption. To overcome this problem, Hedström and Åberg use log linear regression to test the relative importance of social factors and other correlating factors in the probability of unemployed residents of Stockholm gaining employment over a period of time (Hedström 2005). This study suggests that social interactions do determine whether people search for work.

The challenges remaining in this research area are substantial (Ehrlich & Levin 2005). Can we work out how culture is transmitted between individuals and the consequences this has for the development of religion, economics, and our environment? This will require a combination of statistical analysis of correlations between different types of social behavior with an understanding of how ideas are transmitted between individuals.

— Chapter 4 —

Making Decisions

The previous chapter looked at how information is transferred between individuals, in particular when they are looking for food. In one sense we can talk about individuals making decisions about where to collect food. The ants decide which of two food sources to exploit. Under natural conditions, however, food is often depleted or moves. As a result the available alternatives change and it is difficult to define when or if a decision has been made, or to even usefully talk about decisions between alternatives.

There are, however, many situations when animals have to decide between two or more options, whose qualities remain stable through time. This chapter focuses on such situations, where individuals have a number of options and where we can define an end point at which all individuals have made a choice. Most of the examples I consider in this chapter concern how groups choose a new shelter or migrate to a new home. Here we can sensibly talk about decision-making: once all individuals have settled at their new home or shelter, then we can say that a collective decision has been reached. The collective decision-making investigated in this chapter is thus information transfer in a specific, albeit interesting and important, setting. A setting in which there are multiple alternative choices available to a group, and the alternatives remain stable until a point at which we can say a decision has been made.

There are a number of important benefits to an individual in using the information possessed by others in reaching decisions (Sumpter & Pratt 2008). One benefit is the maintenance of cohesion. Choosing the same destination as others, for example, can make an animal less likely to be picked out by a predator. In the search for a new home there are often benefits to consensus, simply because group members do not want to have to invest time and effort re-coalescing because of an initial split.

While information transfer often results in cohesion, the underlying reason individuals followed or copied each other was not necessarily to promote group cohesion. Information transfer can be a form of social parasitism, and cohesion is a disadvantage to the individual who first found the food. There are, however, other potential benefits of information transfer in decision-making. Most importantly, the speed and accuracy of decision-making can both be improved by copying the choice of a better-informed neighbor. Decisions in which cohesion, speed and accuracy are important factors and in which all or nearly all group members come to agree on the same option are often referred to as consensus decisions (Britton et al. 2002; Conradt & Roper 2005). The key question is how individuals reach a rapid consensus for the best of a number of available options.

Consensus Decisions

Cockroaches

Various species of cockroach benefit from increased growth rates when in aggregations (Prokopy & Roitberg 2001). The German cockroach *Blattella germanica* can reduce water loss in dry conditions by clustering together with other cockroaches (Dambach & Goehlen 1999). These cockroaches rest during daytime periods in dark shelters where they aggregate in stable populations (Ishii & Kuwahara 1968; Rivault 1989). These aggregations are at least in part due to attraction to chemical odors on the body of the cockroaches (Rivault et al. 1998). Cockroaches that are collected from different locations and kept isolated as different strains have different odors and are attracted more strongly to the odor of their own strain (Rivault & Cloarec 1998).

Amé et al. (2004) performed symmetry breaking experiments to test the extent of aggregation due to social interactions. Figure 4.1 shows the result of these experiments. Cockroaches were placed in an arena with two identical shelters both with sufficient capacity to shelter all of the cockroaches. In a majority of experiments over 80% of the cockroaches chose the same shelter. These results held even when two different cockroach strains were put in the arena, with different strains usually choosing the same shelter (Amé et al. 2004). Independent of difference in strain, the cockroaches make a consensus decision about which of the two shelters to occupy.

The consensus decision is reached through a very simple rule followed by individual cockroaches: the probability per unit time of an individual leaving a shelter decreases as a function of the number of cockroaches under the shelter. This probability of leaving decreases rapidly as the number of cockroaches under the shelter increases (figure 4.2a). By incorporating this quorum response into differential equation and

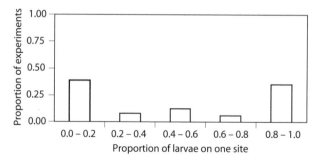

Figure 4.1. Results of Amé's experiments where cockroaches were offered two identical shelters showing the frequency distribution of the proportion of individuals choosing one of the two shelters over 49 trials. (Reprinted by permission from J. M. Amé, C. Rivault, & J. L. Deneubourg, 2004, "Cockroach aggregation based on strain odour recognition," *Animal Behavior* 68:4, 798–801, © Elsevier.)

stochastic simulation models of cockroaches finding and leaving shelters, Amé et al. (2004) showed that it could explain consensus shelter choice. The model they used has strong similarities to the models of ant foraging described in box 3.A. A disproportional response to the presence of other cockroaches is the key to a consensus decision. Amé fitted the function

$$\frac{\theta}{1 + \rho \left(\dfrac{x}{S} \right)^{\alpha}} \tag{4.1}$$

to the probability per second per cockroach of leaving a shelter, where x is the number of cockroaches under the shelter.

The model predicts that if $\alpha = 1$, i.e., the time spent in the shelter is directly proportional to the number of cockroaches under it, then the cockroaches would divide equally between the two shelters. If $\alpha > 1$, i.e., the time spent in the shelter increases more than linearly with the number of cockroaches under the shelter, then symmetry breaking occurs and a consensus is reached for one of the two shelters (Millor et al. 2006). The experimentally measured value of $\alpha \approx 2$ thus accorded with the consensus decisions seen in the earlier experiments (figure 4.1). Further investigation of the model shows that provided that $\alpha > 1$, only a relatively weak positive response to the presence of conspecifics is sufficient to produce symmetry breaking. This could explain why combinations of different strains exhibit an equally strong tendency to make consensus decisions, despite the fact that each strain has only a weak attraction to the odor of the other strain (Leoncini & Rivault 2005).

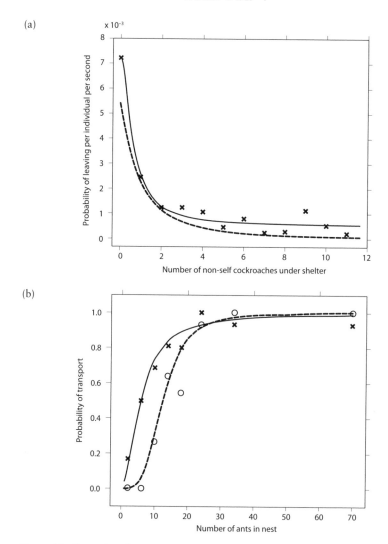

Figure 4.2. Examples of empirical quorum responses in the decisions of migrating insects. (a) Cockroaches. Crosses indicate measured leaving times, dashed line is fit given by Amé et al. (2006) of $\theta/(1+\rho((x-1)/S)^\alpha$, with parameter values $S=40$, $\theta=0.01$, $\rho=1667$ and $\alpha=2$; and solid line is the best fit of the equation $\phi+\theta/(1+\rho((x-1)/S)^\alpha$, with parameter values $S=40$, $\varphi=0.00051$, $\theta=0.0067$, $\rho=1667$ and $\alpha=1.73$. This second fitted line allows for the fact that the probability of leaving does not go to zero with the number under the shelter. (b) A quorum rule governs the probability of a *Temnothorax* scout switching from tandem run recruitment of fellow scouts to faster transport of the bulk of the colony. Crosses show proportions of scouts choosing transport over tandem runs at different populations under high urgency. Open circles show corresponding data under low urgency. Solid and dashed lines, respectively, show a Hill function fit to these data: probability of transport $= x^k/(x^k + T^k)$, where x is the new site population. Reproduced from Pratt (2005b).

Social Insect Migration

As with many other aspects of collective behavior, social insects provide the most detailed experimental studies of decision-making. For many social insects, the survival of the colony depends upon remaining together and making a good decision about where to live. For example, honeybees invest heavily in comb construction, brood-rearing, and food storage at their nest. A poor initial choice of nest site or a failure of all colony members to choose the same nest site can greatly reduce the colony's reproductive success.

Ants of the genus *Temnothorax* live in colonies of between 50 and 500 individuals in small rock or wood cavities. In the laboratory, a colony whose nest has been damaged moves to a new site within a few hours, reliably choosing the best site from as many as five alternatives, discriminating among sites according to cavity area and height, entrance size, and light level (Franks et al. 2003b; Pratt & Pierce 2001). Around 30% of the ants actively partake in the process of choosing a new nest site. These active ants undergo four phases of graded commitment to a particular nest site (Pratt et al. 2005). Each ant begins in an exploration phase during which she searches for nest sites. Once she finds a site she enters an assessment phase, carrying out an independent evaluation of the site, the length of the evaluation being inversely proportional to the quality of the site (Mallon et al. 2001). Once she has accepted the site she enters a canvassing phase, whereby she leads tandem runs, in which a single scout follower is led from the old nest to the new site. These recruited ants then in turn make their own independent assessments of the nest and also recruit once the assessment period is over. The nest population thus increases through a process of recruitment and of independent discoveries of the nest. Since ants take longer to accept lower quality nests, when two alternative nests are presented to the ants recruitment is more rapid to the better quality nest (Pratt et al. 2002).

Recruitment via tandem runs is rather inefficient: ants move at one third of the usual walking speed when leading a tandem run (Pratt et al. 2005). However, rather than indefinitely continuing to recruit using tandem runs, ants recruit in this manner only when the nest population is below a threshold population, referred to as the quorum threshold. Once the population exceeds this quorum threshold a recruiting ant enters a committed phase, where she carries passive adults and brood items to the new nest site (Pratt et al. 2002). These transports are rapid, as carrying another ant does not significantly reduce an ant's walking speed. The meeting of the quorum thus marks a shift from slow to rapid movement into the new nest.

Honeybee emigration usually occurs in spring, when the queen and a swarm of roughly 10,000 worker bees leave their nest and cluster in a densely packed swarm in a nearby tree. Several hundred scout bees then fly from the swarm and search for tree cavities and other potential new homes. Successful scouts use the waggle dance to recruit other scouts to these sites, and those recruited may in turn dance for a site. A positive feedback loop of recruitment to sites begins, similar to that seen when honeybee colonies forage for food (Information Centers, chapter 3). Dances are more frequent for, and thus recruitment is stronger to, better quality sites so the population of recruited scouts grows faster (Seeley & Buhrman 1999; Seeley & Visscher 2004a). Once the number of bees at a site reaches a quorum the bees begin an additional recruitment strategy to dancing, known as piping (Seeley & Visscher 2003; Seeley & Visscher 2004b). Piping is a signal to other non-scout bees at the swarm to warm their flight muscles in preparation for the entire swarm to lift off and fly to the new nest site (Seeley et al. 2003). Over the two or so days during which the scouts search and recruit to new nests, there is dancing for a large number of alternative sites but usually only one site reaches quorum and induces swarm liftoff. Usually at the point of liftoff only one site has reached the quorum threshold population, but in rare cases, split decisions are observed. In these cases the bees lift off in different directions but are then forced to return to the tree branch to begin the process again (Lindauer 1955; Lindauer 1961).

There are strong similarities between the decision processes of *Temnothorax* ants, honeybees, and even cockroaches. All three species exhibit positive feedback and quorum responses (figure 4.2b). The ubiquity of these features and their importance in producing asymmetrical choices (see again box 3.A) suggests that these similarities are, at least in part, an evolutionary consequence of a need by individuals to reach consensus. Nonquorum based recruitment would not give the same degree of consensus.

There are also illuminating differences between species. The ants' and bees' recruitment signals, such as tandem runs, dances, and piping noises, are highly sophisticated. Communication between cockroaches is through attraction to hydrocarbons present on all parts of the cockroaches' bodies (Rivault et al. 1998). Attraction to other cockroaches is thus likely to be cue-based, although the fact that these hydrocarbons differ among strains may indicate that they are an evolved signal to individuals of the same strain (Information Centers, chapter 3). Whether the hydrocarbons are signals or cues, they are certainly a less complicated and probably less costly form of communication than those employed by migrating ants and bees. This greater complexity on the part of the ants and the bees is probably due to a greater requirement for consensus and colony unity. In chapter 9 I discuss how the complex migration algorithms employed by ants and honeybees provide for improved accuracy and an ability to tune decision-making to different environmental conditions.

Other Insects and Spiders

U-shaped distributions, such as that in figure 4.1 for cockroaches and figure 3.1d for ant foraging, are a ubiquitous feature of binary choices by social and gregarious animals (Deneubourg et al. 2002). When offered a T-shaped climbing structure, spiders construct draglines between the bottom and only one side of the two ends at the top of the T-shape (Saffre et al. 2000); when confronted with the choice between ascending from one of two ends of a T-shaped structure, weaver ants build a chain down from only one side (Lioni & Deneubourg 2004); social caterpillars forage on only one of two available branches (Dussutour et al. 2007); and migrating *Messor* ants leave pheromone trails to only one of the available nest sites (Jeanson et al. 2004a)—all show such U-shaped choice distributions. Jeanson et al. (2004b) showed that even "solitary" spiderlings have the same pattern of decision-making when building with draglines. By using the dragline shortcuts provided by the spiders that have already climbed up one side of a Y-shaped cotton thread these spiderlings remain cohesive.

In cases such as these, where individuals modify their environment, it is not entirely clear whether decisions arise because it is advantageous for individuals to be in a group, or because a choice made by one individual alters the environment in a way that makes it easier for other individuals to follow the same path. Thus, while U-shaped distributions are indicative of decision-making in response to the previous actions of others, they should not be necessarily interpreted as resulting from individuals acting in order to promote cohesion.

From a mechanistic viewpoint, U-shaped choice distributions imply a disproportional response to the actions of others. For many of the examples listed above individuals exhibit some form of quorum threshold, similar to those seen in the migrating *Temnothorax* ants and cockroaches. Either the probability of leaving a group decreases as a sharply non-linear function of the number of members (figure 4.2a), or the probability of joining a group increases as a sharply non-linear function of the number of members (figure 4.2b). These empirical observations demonstrate a basic property of all collective decision-making and information transfer: positive feedback together with quorum responses lead to U-shaped choice distributions.

Evolutionarily Stable Decisions

There are many situations where it is beneficial for individuals to reach consensus, but the number of individuals that can take a particular action is limited. Again looking at how cockroaches divided themselves between two shelters, Amé et al. (2006) investigated how the capacity of shelters affected the distribution of cockroaches between them. They found that when the capacity of a single shelter was insufficient to house all the cockroaches, the cockroaches split 50:50 between the two shelters, but when the capacity of both shelters was sufficient for all then the majority would choose the same shelter. The split in this case was nearer to 80:20 or 20:80. Furthermore, the switch from 50:50 to 80:20 occurred when both shelters had almost exactly the capacity to house all of the cockroaches (figure 4.3a). These results were consistent with predictions of their earlier mechanistic model of cockroach aggregation, with equation 4.1 playing a central role in determining that a majority choose the same shelter (figure 4.3b).

Based on these experimental results and earlier studies of the advantages of aggregation, Amé et al. (2006) suggested an optimality model for shelter choice. They proposed that the benefit of being in a shelter increases at first with number under the shelter, but as the capacity of the shelter is reached this benefit decreases due to overcrowding and the possibility of being exposed on the edge of the shelter. Box 4.A describes how the fitness function arising from their analysis can be used to determine both the optimal group size and the evolutionarily stable group size as a function of shelter capacity (see chapter 2 for a discussion of optimal and stable group sizes).

Amé et al. (2006) looked only at the optimal group size (figure 4.3c) and concluded that the mechanisms employed by the cockroaches and the experimental data supported the hypothesis that the cockroaches are able to make optimal decisions. This result is surprising from the viewpoint of individuals maximizing their own fitness, since the stable group size gives a quantitatively different prediction (figure 4.3d). The optimal group size model predicts that the distribution of cockroaches between shelters will bifurcate from 50:50 to being biased towards one of the two shelters at roughly $S = N$, i.e., when the size of each shelter equals the number of cockroaches. The stable group size model predicts that the switch to a biased distribution will occur while $S < N$. Even when each shelter is too small to house all cockroaches, it is often beneficial for an individual cockroach to aggregate with the majority. Thus if each cockroach tries to maximize its own fitness, the average cockroach does worse than were it to aggregate according to the optimal group size.

Theoretical comparison of optimal and stable group size raises a number of interesting questions. Assuming that cockroaches are not highly related to each other and/or are competing locally for resources, we would expect them to adopt the evolutionarily stable rather than the optimal strategy in shelter choice (see chapter 10). Despite this, at first sight it is the optimal model (figure 4.3c) that gives the best fit to the data (figure 4.3a) and best matches the mechanistic model (figure 4.3b). However, the data is a distribution of experimental outcomes and it is difficult to be sure at exactly which point the bifurcation occurs. Different parameterizations of the models give different bifurcation points and tuning the parameters, none of which have been measured for the fitness function, can give quite different quantitative predictions. The experimental setup, where cockroaches are placed in an unfamiliar arena, is different from their natural environment, where they also learn about and use navigational cues in finding shelter (Durier & Rivault 1999). More detailed studies are needed to find out exactly what individual cockroaches might be trying to optimize.

Despite these potential problems, Amé's results are powerful because of their simultaneous study of experimental, mechanistic, and functional explanations within a single system. Theoretical studies of evolutionarily stable strategies for choosing resource sites have also found bifurcations in site choice. For example, Moody et al. (1996) model individuals that can choose between two patches with the aim of maximizing food intake rate while minimizing predation risk. Both food intake and predation risk decrease with the number of conspecifics at the patch, but with differing functional forms. They found that as a predator dilution factor increased, i.e., there were greater benefits to cohesion, a bifurcation occurred from a relatively even distribution of individuals between patches to individuals aggregating at one of two patches. This study, based on a functional explanation of patch choice, shows that even if the exact functional form of Amé's cost/benefit function is not correct, the mechanism adopted by cockroaches in decision-making is consistent with the principle of individuals making decisions that attempt to increase their fitness.

As with many aspects of information transfer in groups, an exciting challenge in this area is linking mechanistic and functional approaches. Jackson et al. (2006) have begun to address this challenge by looking at how natural selection might act on mechanistic rules for how individuals change their vigilance behavior and decide to migrate between patches when confronted with a predation risk. These models again predict bifurcations and U-shaped distributions of numbers of individuals choosing different patches. Indeed, many of the predictions are highly reminiscent of the predictions first arising from purely mechanistic models (Camazine et al. 2001). This would suggest that the models of Jackson and others could benefit by incorporating the response thresholds, emphasized by Deneubourg, Amé, and colleagues, into evolutionary models of decision-making.

(a)

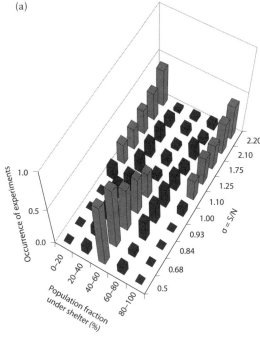

(b)

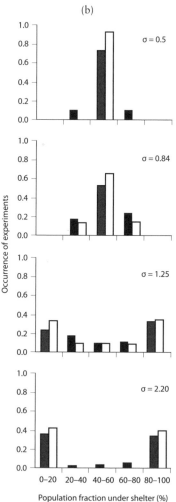

Figure 4.3. Distribution of cockroaches among shelters for different shelter size. (Reproduced from J.M. Amé, J. Halloy, C. Rivault, C. Detrain, & J. L. Deneubourg, 2006, "Collegial decision making based on social amplification leads to optimal group formation," *PNAS* 103, 5835–5840, figs. 1b & 2b, © 2006 National Academy of Sciences, USA) (a) Experimental results; (b) predictions of Amé's mechanistic model, white bars show model prediction while dark bars show experimental outcome (reproduced from Amé et al. 2006); (c) predictions of optimal group size model (see box 4.A); and (d) predictions of stable group size model (again see box 4.A).

(c)

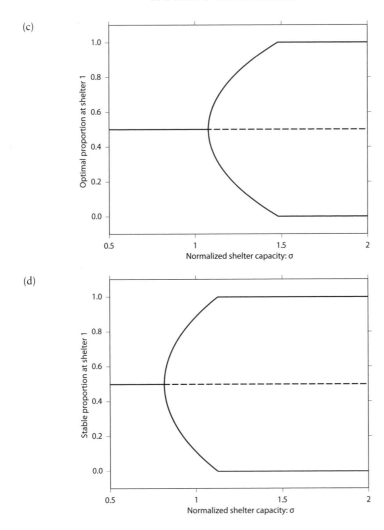

(d)

Many Wrongs

In the last chapter I looked at situations where one individual has a piece of information, such as the location of some food, which is transferred to others through positive feedback. In such situations it is often clear that there is an advantage to copying the behavior of the individual with the information. There are, however, many decision-making situations when a group of individuals are faced with two or more options, with none of them having more information than the rest about which of the options is

Box 4.A Optimal and Stable Cockroach Distributions

Amé et al (2006) proposed the following fitness function for a cockroach under a shelter where a fraction x of the other cockroaches are under the same shelter:

$$f(x) = (1 + p(x/\sigma)^2)(1 - x/\sigma),$$

where p is a constant and $\sigma = S/n$, where S is the size (or capacity) and n is the total number of cockroaches in the arena. Group fitness, i.e., the average fitness per individual, is $xf(x) - (1 - x)f(1 - x)$.

To find the maximum of this expression, we differentiate it and solve after setting equal to zero. This gives three solutions at $x = 1/2$ and

$$x = \frac{p \pm \sqrt{-3p^2 - 2p\sigma^2 + 3\sigma p^2}}{2p}.$$

The solutions corresponding to maxima are shown in figure 4.3c as a function of the normalized shelter capacity σ. At a critical value of σ group fitness goes from having a single maximum at $x = 1/2$ to having two maxima corresponding to an aggregation in one of the two shelters. The exact value of this critical point is determined by the value of p, but for a wide range of these values the critical point is slightly larger than $\sigma = 1$.

As we saw in chapter 2, and is further illustrated in a number of examples in chapter 10, the strategy or behavior that is optimal for the group is not necessarily that which is stable with respect to individuals attempting to maximize their own fitness. For example, assume that $n = 10$, $\sigma = 1$ and $p = 10$. It is then optimal for the group to split 5:5 between the two shelters. In this case $f(5/10) = 1.75$ but $f(5/10 + 1/10) = 1.84$. Thus if one of the five individuals at one shelter moves to another shelter it will increase its fitness, while simultaneously decreasing the fitness of those in the shelter it leaves as well as the average fitness of all individuals in the population. The group optimal distribution between shelters is not the stable distribution between the shelters. Individuals trying to maximize their own fitness will not choose the group optimal distribution.

The stable distribution of cockroaches among shelters corresponds to the values of x at which $f(x) = f(1 - x)$. At this proportion it is no longer beneficial for individuals to move to another shelter. Solving $f(x) = f(1 - x)$ gives

$$x = \frac{p \pm \sqrt{-3p^2 - 4p\sigma^2 + 4\sigma p^2}}{2p}.$$

Provided $p > 2\sigma$ then there will exist carrying capacities where an equal division between the two shelters is optimal but not stable. Figure 4.3c and 4.3d show an example where an equal division between shelters becomes unstable when $\sigma = 0.815$, but is optimal for the group up until the point that $\sigma = 1.017$. In general, a range of carrying capacities exists where the stable and optimal proportions at the shelter differ.

best. For example, in an unfamiliar environment individuals must choose where to look for food. In such cases, each individual has some probability of making the "correct" decision, but no individual is a priori more likely to be correct than any other.

In binary choices, we can assume that each individual has a probability p of making a correct decision in the absence of others with which to confer. Figure 4.4 shows the probability that the majority make a correct decision, provided each individual makes its decision independently of the others (see box 4.B for details). Although the probability that each individual is correct is only $p = 0.6$, the probability that the majority of the group is correct increases steeply with group size. Groups of size 100 will hardly ever make a majority error. This result was first applied in the 18th century by Condorcet to designing the jury system. In general, it illustrates that majority decisions are good at pooling information and improve decision-making accuracy (King & Cowlishaw 2007; List 2004).

A related concept is that of many wrongs. For example, navigating animals possess directional information—from visual landmarks, internal compass, smell and so on—that is subject to error. Assuming this error is unbiased, then the average direction of the group is more likely to be correct than that adopted by one randomly chosen individual (Simons 2004; Wallraff 1978). This argument can be formalized as an application of the central limit theorem, which predicts that the error in the average direction decreases in proportion to the square root of the group size. Experimental tests of this theory on navigating birds have had mixed results, but do appear to show some increase in accuracy with group size (Biro et al. 2006;

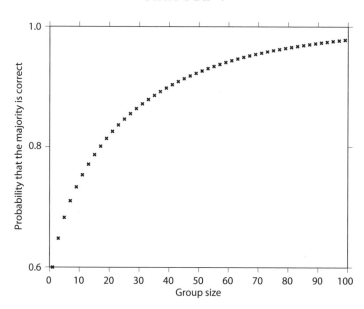

Figure 4.4. Condorcet's theorem. The probability that the majority of individuals are correct (for odd numbers of individuals) when each is correct with probability $p = 0.6$.

Simons 2004; Tamm 1980). Oldroyd et al. (2008) looked at the dances of *Apis florea* honeybees within a swarm prior to liftoff. These dances encode the direction of proposed nest sites. They found that the actual direction taken by the swarm was very close to that of the average direction indicated by the dances. This would suggest that the dancing bees can effectively integrate their directional information and lead a large group of uniformed individuals in the average of their proposed directions.

In humans, the "many wrongs" principle is highlighted by an observation by Galton (1907). He examined 800 entries into a "guess the weight of the ox competition," where a crowd of fairgoers each paid a small amount to guess how much a large ox would weigh after slaughter, with the most accurate guess winning a prize. Although the guesses had a wide variation the average guess was only 1 pound (450 g) less than the 1197 pounds (544.5 kg) that the ox weighed. Acting independently, the crowd "knew" the weight of the ox. There are many such examples of collective accuracy in humans: ask the audience on "Who wants to be a millionaire?" ; the accurate prediction of American presidential elections by betting; and the Google search engine using links to a webpage to measure its popularity are just some (Surowiecki 2004).

Variability is an inherent feature of animal groups, including insect societies (Jeanne 1988; Seeley 1995). From only one-week old, before they

Box 4.B Condorcet's Theorem and the Central Limit Theorem

Condorcet's theorem is as follows. Assume that an odd number of individuals n have to make a choice between two options independently of the others and each has a probability p of being correct. The probability that the majority make the correct choice follows directly from the derivation of the binomial distribution, i.e.,

$$m(n, p) = \sum_{i=\frac{n+1}{2}}^{n} \binom{n}{i} p^i (1-p)^{n-i}.$$

Figure 4.4 plots this function for $p = 0.6$. As the number of individuals goes to infinity, $m(n, p) \to 1$ and the majority decision is always correct. In the case where n is an even number we have to make a choice about how we treat cases where an equal number make the same choice, but the overall shape of the curve remains the same.

The principle of many wrongs assumes that there are n independent individuals each of which attempts to estimate some continuous variable that has a correct value v. We let X_i be the estimate of the value by individual i, such that the expectation is unbiased, i.e., $E[X_i] = v$, and the variance in the $Var[X_i] = \sigma^2$ estimate is the same for all individuals. The X_i can be distributed according to any distribution with finite variance. Figure 4.5 shows the distribution of average estimates of $n = 1$, 5, and 25 individuals under various estimate distributions. For distributions with large variance, such as the Bernoulli and lognormal, the estimate of a single individual often lies a long way from the correct value v. Indeed, individuals making estimates according to a Bernoulli distribution never estimate correctly. However, even if the individual estimates are widely scattered the average of 25 and even only 5 estimates lies much closer to the true value. As in Condorcet's theorem, the more individuals that we average over, the closer we come to the true value.

The central limit theorem states that for large n the mean estimate, $1/n \sum_{i=1}^{n} X_i$, is distributed normally with mean v and variance σ^2/n. This theorem gives us three useful pieces of information: (1) that independent of the distribution of estimates by a single individual, the mean estimate becomes normally distributed; (2) the degree of

(Box 4.B continued on next page)

error in the average of n estimates has standard deviation $\sigma/\sqrt{n}$; and (3) for very large n this error tends to zero. Figure 4.5 shows the frequency of average estimates predicted by the central limit in comparison with the actual distribution. By $n = 25$ this mean estimate is very close to the correct value.

Although the many wrongs principle seems to provide a powerful way for groups to make correct decisions it relies on two key assumptions, that individuals are independent and that they are unbiased. In the main text I discuss the various problems with the assumption of independence. The assumption of lack of bias must also be treated with care. For example, if we set up a navigation experiment that deliberately misleads subjects, then no matter how many subjects we independently examine they will each be misled and thus choose a biased and less accurate route. Determining variation due to error and variation due to internal bias thus poses a difficult problem in practice.

have left the hive for the first time, honeybees have different levels of response to sucrose, which later in life determines their propensity to collect water, nectar, and pollen (Pankiw & Page 2000). This variability can lead to benefits for the colony. For example, genetically diverse honeybee colonies keep a more constant brood nest temperature than genetically uniform ones (Jones et al. 2004; Mattila & Seeley 2007). Stability is thought to be maintained because of individual differences in the temperatures at which individuals begin and stop fanning (Graham et al. 2006; Jones et al. 2007; Weidenmuller 2004). Individuals responding at different temperatures avoid all or nothing responses that could lead to the colony overshooting its target temperature (Sumpter & Broomhead 2000).

Integrating Many Wrongs

The assumption that individuals are independent leads to a paradox in the theory of many wrongs. On the one hand the theory says that the group is collectively wise, but on the other hand it requires individuals to be independent. If there is too much conferring among individuals before they reach a final decision then their decisions are no longer independent. Positive feedback can spread particular information quickly through the group, encouraging all individuals to make the same, possibly incorrect,

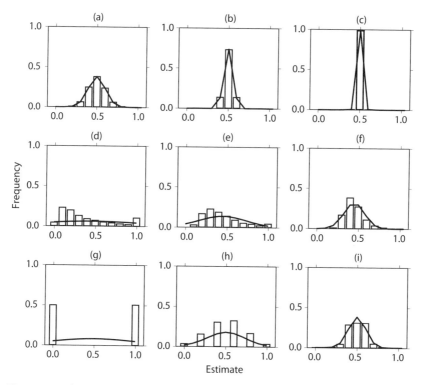

Figure 4.5. The many wrongs principle and the central limit theorem. The distribution of average estimates when estimates are distributed according to: (a–c) normal distribution with $v = 0.5$ and $\sigma = 0.1$; (d–f) lognormal distribution with $v = 0.5$ and $\sigma = 0.58$; and (g–i) a Bernoulli distribution with $v = 0.5$ and $\sigma = 0.5$. The histograms give the frequency distribution for: (a, d, & g) the estimate of a single $n = 1$ individual; (b, e, & h) the average estimate of $n = 5$ individuals; and (c, f, & i) the average estimate of $n = 25$ individuals. The solid lines overlaid on the histogram give the prediction for the distribution of estimates by the central limit theorem, i.e., the predicted frequency for a normal distribution with mean $v = 0.5$ and variance σ^2/n.

choice. Alternatively, if there is too little conferring then each individual will act independently and fail to benefit from the input of others.

In human decision-making, in situations where all individuals are agreed on the best outcome but are individually unsure about the best course of action to secure this outcome, the many wrongs paradox lies at the basis of the phenomena of "groupthink" (Janis 1972, 1982). Groupthink is where pressures of group members on each other lead to a narrowing down of opinions. It is most likely to occur in groups where members have similar backgrounds and interests. Janis (1972) proposed that groupthink can be prevented by allowing a large number of individuals to first collect information independently before presenting their recommended course of action to a smaller number of centralized evaluators. By correctly weighting the information presented by the independent individuals, which is itself no easy task, the evaluators can then make a decision based on an average of the opinions presented.

While human groups may be able to integrate complex information from a large number of sources when making decisions, this is not always possible for animal groups. In most cases decision-making by animal groups is decentralized (Seeley 1995, 2002) and as such positive feedback plays a necessary role in their decision-making (Bonabeau et al. 1997; Deneubourg & Goss 1989). As we saw in the last chapter, although not always leading to the correct choice, positive feedback through pheromone trails usually allowed ants to choose the best of two available food sources. Similarly, not only do *Temnothorax* ants and honeybees make consensus decisions, but they are also able to choose the best of a number of alternatives (Mallon et al. 2001; Seeley & Buhrman 2001).

Box 4.C investigates how positive feedback combined with quorum responses can aid accuracy in decision-making by groups without full consultation of all group members. The model assumes very limited cognitive powers on the part of individuals. In particular, they have no way of directly comparing the two available options. Instead the probability of choosing an option is simply an increasing function of the number already there. The question is what functional form of response gives the most accurate decisions? The model shows that accurate decision-making is achieved with quorum-like responses at a fixed threshold, rather than smooth linear responses (figures 4.6 and 4.7). This modeling result suggests that response thresholds not only provide cohesion, but also facilitate accuracy (Sumpter & Pratt 2008).

While the quorum mechanism leads to some improvement in accuracy over individual decisions, it does not achieve the level predicted by Condorcet's theorem. Indeed, Condorcet's theorem provides an upper bound for the accuracy of collective decision-making. For example, given $N = 40$ individuals, each with a 1/3 probability of making the wrong choice, then by Condorcet's theorem, the probability of a majority error is just 3.33%. This is lower than even the most accurate decisions made using quorum responses: for steep thresholds of between 5 and 15 and low spontaneous accept rates, approximately 10% of individuals take the least favorable option. Despite not reaching the upper bound for accuracy, a simple copying rule based on threshold responses substantially reduces errors compared to purely independent decision-making.

To test both decision-making mechanisms and the extent to which decision-making improves with group size, Ward et al. (2008) investigated how fish make movement decisions in response to others. They

Box 4.C Quorum Responses and Decision-making Accuracy

Consider a group of n individuals initially uncommitted to either of two available options. Each of these finds one of the two options with a constant probability r per time step. This probability is independent of the actions of others. If an individual arrives at an option and no one else is there, then he or she commits to it with probability ap_x for option X and ap_y for option Y. If an individual arrives at an option and other individuals are present, the probability of committing and remaining at the option is an increasing function of the number already committed. Specifically, if x is the committed number at the option then the probability that the arriving individual commits is

$$p_x\left(a + (m - a)\frac{x^\alpha}{T^\alpha + x^\alpha}\right), \qquad (4.C.1)$$

where a and m are respectively the minimum and maximum probability of committing; T is the quorum threshold at which this probability is halfway between a and m; and α determines the steepness of the function. Equation 4.C.1 encodes a range of possible responses to the number that have chosen a particular option. In particular, as α increases the response approaches a step-like switch, or quorum response, at the threshold T. In the model, a similar function determines the probability of selecting option Y, and by setting $p_x > p_y$, we assume that individuals prefer X to Y.

Figures 4.6a and 4.6b give examples of the choices over time of $n = 40$ individuals for shallow proportional responses ($T = 10$ and $\alpha = 1$) and steep quorum responses ($T = 10$ and $\alpha = 9$), respectively. For both types of responses, the proportion of committed individuals grows slowly for the two options, but slightly faster for the preferred option X. After the number of adherents to X reaches the threshold T, commitment to X significantly outpaces commitment to Y. Averaged over 1000 simulations, 75.5% of individuals choose X for a shallow response, while 83.3% do so for the steep quorum response. In both cases the proportion choosing the better option is higher than were each to make an independent decision, in which case $p_x/(p_x + p_y) = 66.7\%$ would be expected to choose X. Thus, in these simulations choices based on copying others

(Box 4.C continued on next page)

reduce individual errors and make group decision-making more accurate than independent assessment alone. While a steep quorum response led on average to more accurate decisions, the distribution of decision-making accuracy is wider for $\alpha = 9$ than for $\alpha = 1$ (figures 4.6c and 4.6d). This observation reflects the amplification of small initial errors for steep responses. If through random fluctuations, the least favorable option happens to be chosen by more than a threshold number of individuals, then the quorum rule amplifies these early errors and nearly all individuals make the same incorrect choice.

Decision-makers typically face a trade-off between speed and accuracy. In the simulations, a steep quorum function, $\alpha = 9$, yielded a more accurate decision, but the time taken for all individuals to choose was longer on average (307.8 ± 71.0 time steps, mean ± standard deviation) than when $\alpha = 1$ (253.7 ± 64.0 time steps). In order to investigate how different values for α, T, and a affect speed and accuracy, the parameters are systematically varied and their effect on the time needed for all individuals to make a choice and the proportion choosing the better option are measured (figure 4.7). The results show that speed is maximized by setting a to its maximum value of 1 (assuming that $m = 1$ as well). Greater speed, however, comes at the expense of more individuals choosing the worse option. Accuracy is maximized with low a, high α, and T of around 10, but these values also produce relatively slow emigrations. For a more detailed analysis of this model see Sumpter & Pratt (2008).

presented small groups of three-spine sticklebacks with a Y-shaped maze. A drag line was set up down either side of the maze, along which replica conspecifics were drawn. The proportion of individuals following to either side of the maze was plotted as a function of the number of replicas traveling in each direction and the group size. The sticklebacks tend to follow the replica, with smaller groups (1 or 2 fish) more likely to be influenced by the replicas than large groups (4 or 8 fish). If the difference between the number of replicas going to the two sides was only 1 (e.g., if left:right was 1:0 or 2:1) the larger groups were not influenced by the majority. If the majority was 2 however (e.g., if left:right was 2:0 or 3:1) then the larger groups were much more likely to follow the majority.

A model similar to that presented in box 4.C fit the data. The main difference in this case is the inclusion of the number of undecided fish in an

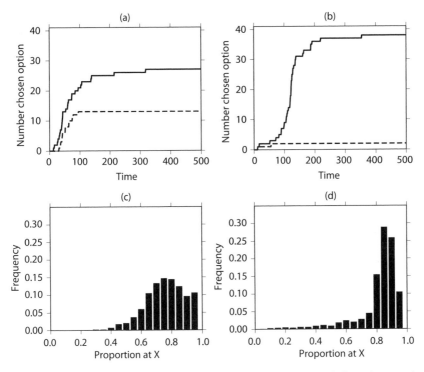

Figure 4.6. Simulations of a simple quorum response model, for (a, c) shallow ($k = 1$) and (b, d) steep ($k = 9$) thresholds. Both (a) and (b) plot the change in the number of individuals committed to options X (solid line) and Y (dotted line) for one simulation with $k = 1$ and $k = 9$, respectively; (c) and (d) show the distribution of the proportion of individuals choosing X after everyone has decided. Other parameters are $r = 0.02$, $p_x = 1$, $p_y = 0.5$, $T = 10$, $a = 0.1$, and $m = 0.9$. Reproduced from Sumpter & Pratt (2008).

individual's decision to go left or right. In particular, we assumed that the probability of an individual going left on time step $t + 1$ is

$$a + (m - a)\frac{(L(t) - L(t - \tau))^\alpha}{U(t)^\alpha + (L(t) - L(t - \tau))^\alpha + (R(t) - R(t - \tau))^\alpha}, \quad (4.2)$$

where m is the maximum probability of committing to a decision; $U(t)$ is the number of uncommitted individuals at time t; and $L(t)$ and $R(t)$ are, respectively, the total number of individuals that have gone left and right by time t. Three parameters—a, which is the spontaneous accept rate; α, which is the steepness of response; and τ, which is the number of time steps over which fish are influenced by individuals that have already made decisions—determine the shape of this response (Ward et al. 2008).

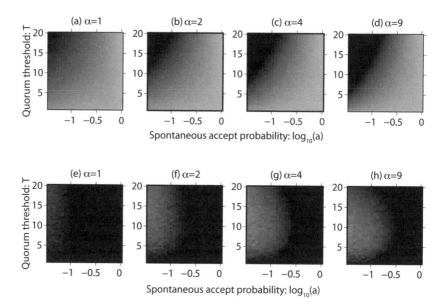

Figure 4.7. Speed and accuracy of decision-making for the simple quorum response model. Predicted effects of the parameters a, T, and k on (a–d) the time until all individuals have made a decision and (e–h) the accuracy of that decision. Darker shading corresponds to slower and less accurate decisions. In each image, a and T are varied for different threshold steepness, k. The plots show mean duration (time steps of the model) and accuracy (proportion of individuals choosing the less attractive option Y) over 1000 simulations for each parameter combination. Reproduced from Sumpter & Pratt (2008).

The parameters of this model were measured from the experiments and it was found that a steep quorum-like response ($\alpha \approx 3$) gave the best fit to the data. It appears that in deciding whether to go left or right the fish weigh the numbers going in each direction and their current group size and are disproportionately likely to take the direction of the majority.

Further experiments on sticklebacks looked at the likelihood of a group following a leader into a potentially dangerous situation. Groups of 4 or 8 fish swam past a predator replica only when guided by 2 or more replicas, whereas single replicas were mostly ignored in this situation. On the other hand, single individuals could be fooled into following a single replica past a predator that they would almost never approach when alone. Interpreting these results in terms of our model we see that uncommitted individuals in larger groups only follow above a threshold number of leaders. This threshold dramatically reduces the probability of errors being amplified throughout a group because if the probability one individual makes an error is small, say ε, then the probability that two fish independently make errors at the same time becomes very small, i.e.,

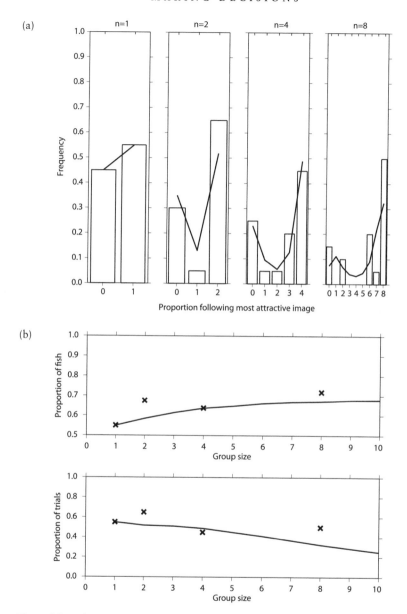

Figure 4.8. Fish movement decisions in response to different replicas. In the experiments a large (more attractive) replica fish moved in one direction and a standard-sized fish moved in the other direction. (a) Comparison of data (histogram) and model (solid lines) in terms of proportion of fish following the most attractive replica. (b) Proportion of fish making the "more attractive" choice (top panel) and proportion of trials in which all fish make the "more attractive" choice (bottom panel) for the data (crosses) compared to the average of 1000 simulations of the quorum-response model (solid line). See Sumpter et al. (2008b) for details.

ε^2. Interestingly, this rule of following only when two other individuals make a particular choice is consistent with experiments where humans are asked to make a decision after hearing the opinions of others (chapter 3, figure 3.9).

The quorum rule allows fish to make more accurate decisions as group size increases. Sumpter et al. (2008b) looked at how well groups of fish could discern between two replica leader fish as a function of group size. Figure 4.8a shows the results of an experiment in which a group observed one large replica fish move in one direction and one standard sized fish move in another direction. As group size increases more of the fish follow the larger, more attractive fish. Similar results were observed in 9 other trait comparisons (e.g., fat vs. thin, dark vs. light, etc.). These results were reproduced by a model based on equation 4.2. The model predicted that the frequency with which fish choose the correct option would increase with group size, while the frequency with which all fish would choose the correct option would decrease would group size. This prediction was confirmed in the data (figure 4.8b). The data also confirm Condorcet's prediction that decision-making improves with group size (Sumpter et al. 2008b).

As with the cockroaches, ants, honeybees, and humans (see Quorum Mechanisms for Information Transfer, chapter 3), we see that positive feedback and quorum responses are a key mechanism in fish decision-making. Measuring the form of these responses across species will further help determine the importance of information transfer in the evolution of group-living in these species. Interestingly, these steep threshold responses can sometimes amplify random fluctuations and lead to mass adoption of incorrect choices (Sumpter et al. 2008b). This sort of process may account for observations of mass copying (Dall et al. 2005; Laland 1998) or peer pressure in humans (Milgram 1992; Milgram et al. 1969), and may lead animals in groups to make decisions they would not make by themselves. Although quorum responses lead to poor decisions in some notable cases, on average they allow greater accuracy than do complete independence or weak responses to the behavior of others. Quorum responses allow effective averaging of information without the need of complex comparison between options.

— Chapter 5 —

Moving Together

$\mathcal{S}$ome of the most mesmerizing examples of collective behavior are seen overhead every day. V-shaped formations of migrating geese, starlings dancing in the evening sky, and hungry seagulls swarming over a fish market, are just some of the wide variety of shapes formed by bird flocks. Fish schools also come in many different shapes and sizes: stationary swarms; predator avoiding vacuoles and flash expansions; hourglasses and vortices; highly aligned cruising parabolas, herds, and balls. These dynamic spatial patterns often provide the examples that first come into our heads when we think of animal groups.

While the preceding three chapters described the dynamics of animal groups, they did not explicitly describe the spatial patterns generated by these groups. For example, the decision-making of insects and fish was studied in situations where individuals have only two or a small number of alternative sites to choose between. In models of these phenomena, space is represented as the number of individuals who have taken each of these alternatives. This approach often simplifies our understanding of the underlying dynamics of these groups, but in doing so it can fail to capture the spatial structure that characterizes them. As a simple consequence of the fact that these groups move, we need to give careful consideration to how they change position in space as well as time.

The main tool I will use in describing the dynamics of flocking are self-propelled particle (SPP) models (Czirok & Vicsek 2000; Okubo 1986; Vicsek et al. 1995). In SPP models "particles" move in a one-, two-, or three-dimensional space. Each particle has a local interaction zone within which it responds to other particles. The exact form of this interaction varies between models but typically, individuals are repulsed by, attracted to, and/or align with other individuals within one or more different zones. These models allow us to investigate the conditions under which collective patterns are produced by spatially local interactions.

Attraction

Before animals can create spatial patterns they must first come together. In chapter 2, I discussed how and why animal groups form without specific reference to spatial structure. A good starting point for explicitly representing space comes from Niwa (2004). His model, which is an extension of a non-spatial model described in chapter 2, describes groups of individuals that are constrained to move on a lattice (see box 5.A). Each group performs a random walk and when groups meet they merge. Groups split with a fixed probability per time step. Figure 5.1a shows an example of how composition of these groups changes through time and space. Over time groups "clump" together. Sites containing large groups are usually located near to other sites containing large groups, while sites with few individuals are surrounded by other sites with few individuals. The position of these clumps changes through time as the groups move according to a random walk.

The unit of description in Niwa's model is the group. The model defines rules for how groups merge and split. The strength of this approach is that it reproduces the empirical distribution of fish school sizes (compare figure 5.1b and figure 2.6). The main limitation of this model is that it does not describe how between-individual interactions produce group dynamics. Establishing such a connection is often the central question in the study of flocking. It is here that self-propelled particle models play an important role.

In the simplest SPP model the only interaction between individual "particles" is attraction (box 5.B). Figure 5.2a shows the outcome of a one-dimensional SPP model in which individuals are attracted to other individuals within a fixed distance. As in Niwa's model, relatively stable clusters of individuals quickly form. Unlike Niwa's model, larger clusters move slower than solitary individuals. This is because individuals on the edge of the cluster are attracted inwards, resulting in a constant pull towards the center of the cluster's mass. As clusters increase in size they move less and less, while solitary individuals and smaller groups move and eventually join the clusters (Okubo 1986). After some time a small number of large stationary clusters form.

Such aggregation dynamics are seen in cockroach groups (Jeanson et al. 2005). Cockroaches interact via antennal contact and are attracted to other cockroaches through physical contact. Thus, relative to the size of their environment, their zone of attraction is small. Jeanson et al. (2005) placed small groups of cockroaches in a circular arena and watched their aggregation behavior. Since cockroaches are strongly attracted to walls, most of their movement is constrained to the edge of this arena. In effect,

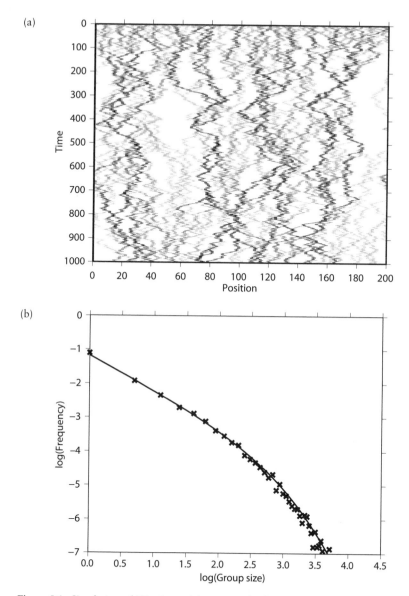

Figure 5.1. Simulation of Niwa's spatial merge and split model. Simulation of model described in box 5.A with $s = m = 200$ sites/individuals and split probability $p = 0.05$. Initially each site contains a single individual, i.e., a group of size 1. (a) Time evolution of the number of individuals across the sites. Darker shading indicates larger groups at a particular site; white indicates sites containing no individuals. (b) Distribution of the number of individuals in a randomly chosen site over 100,000 simulation time steps. The solid line is equation 2.1 with $\langle N \rangle_P$ estimated directly from the simulation.

Box 5.A Niwa's Spatial Merge and Split Model

The basic assumptions of this model are the same as in box 2.B. A total of m individuals are initially randomly distributed across s sites, and n_i represents the number of individuals on site i. The key difference in the spatial model is how the groups move. Here we assume that groups move on a d-dimensional lattice of discrete sites, such that each site has $2d$ neighboring sites, e.g., in one dimension each site has neighbors to the left and right and in two dimensions each site has neighbors to the north, east, south, and west. The lattice is structured so that individuals moving off, for example, the north edge of the lattice reappear at the south. Thus the lattice is a circle in one dimension and a torus in two dimensions. On each time step, each group either moves to one of the neighboring sites, each chosen with equal probability $1/(2d)$; or with probability p the group splits into two groups, one that stays on the same site and the other that moves to a randomly chosen neighboring site. When a group splits the size of the two components is chosen uniformly at random, so that all group sizes are equally likely. If two groups of size n_i and n_j meet at site k, then they form a new group $n_k = n_i + n_j$. Thus, groups always merge when they meet. The same rule applies if three or more groups meet.

Figure 5.1a shows a simulation of the above model in one dimension ($d = 1$). From an initial distribution where each individual occupies one site, larger groups quickly form. These groups perform a random walk and increase in size as they meet other groups. After 1000 time steps there are around five or six large groups and a number of smaller groups. Figure 5.1b shows the distribution of group sizes at a randomly chosen site over 100,000 time steps of the simulation. Niwa (2004) went on to show that the distribution of group sizes in these simulations is characterized by the same curve as in his earlier non-spatial model (box 2.B). By finding the mean group size experienced by an individual it is possible to give an expression for the entire distribution of group sizes.

the attraction to the arena edge means that movements of the cockroaches take place in one dimension and the aggregation process can be visualized by plotting the angular position of the cockroaches through time (figure 5.2b). In experiments where cockroaches were initially placed at random within the arena, a cluster quickly formed containing nearly all

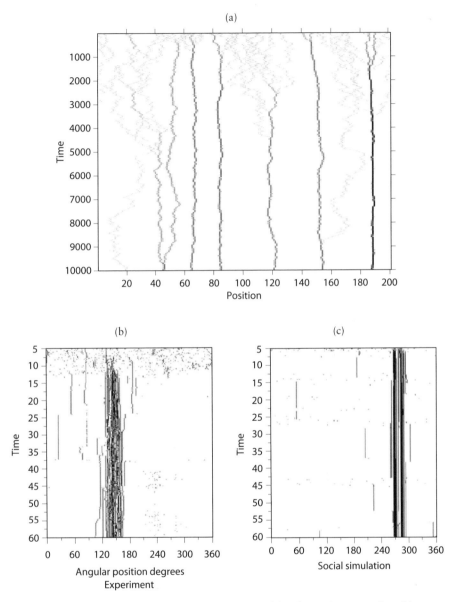

Figure 5.2. Outcome of (a) the simple attraction model in box 5.B compared to (b) experiments on cockroach aggregation (reproduced from R. Jeanson, C. Rivault, J. L. Deneubourg, S. Blanco, R. Fournier, C. Jost, & G. Theraulaz, 2005, "Self-organized aggregation in cockroaches," Animal Behaviour 69:1, 169–180, fig. 4b, © Elsevier), and (c) Jeanson et al.'s detailed individual-based model (reproduced from Jeanson et al. 2005, fig. 4c).

Box 5.B Self-propelled Particle Models

The term self-propelled particle (SPP) was introduced by Vicsek et al. (1995), but the idea of building models where individuals interact through zones of repulsion, attraction, and alignment had been proposed independently by a number of authors (Aoki 1982; Gueron et al. 1996; Helbing & Molnár 1995; Okubo 1986; Reynolds 1987). This box presents some of the simplest of these models, including a model of aggregation and Vicsek et al.'s original SPP model of alignment, as well as a more detailed model by Couzin et al. (2002) including repulsion, attraction, and alignment.

The general SPP model involves a group of N particles in a d-dimensional space. Let the vectors $\underline{x}_i$ and $\underline{u}_i$ represent the position and velocity of individual i. Let r represent the interaction radius of the individuals. On each time step t, all individuals update their position and velocity as follows

$$\underline{x}_i(t+1) = \underline{x}_i(t) + v_0 \underline{u}_i(t+1)$$
$$\underline{u}_i(t+1) = \alpha \underline{u}_i(t) + (1-\alpha)\underline{s} + \underline{e},$$

where v_0 is a constant determining a baseline distance that individuals move per time step, and α is the inertia of an individual (i.e., its tendency to keep the same direction as on the previous time step). The vectors $\underline{s}$ and $\underline{e}$ are determined on each time step for each individual: $\underline{s}$ is a vector (usually a unit vector) with a direction that depends on the position and velocity of the set of particles R_i, which are within distance r of the individual, excluding itself; $\underline{e}$ is a random vector incorporating noise into the movement of the individual and may also be a function of the position and velocity of i's neighbors.

Attraction: To model individuals that are attracted to one another the vector $\underline{s}$ should point toward the average position of an individual's neighbors. In one dimension we can set

$$s = \frac{1}{|R_i|} \sum_{j \in R_i} sign\{x_j(t) - x_i(t)\}.$$

The function $sign\{a\}$ returns 1 if $a > 1$, -1 if $a < 1$, and 0 if $a = 0$. We set e to be a number selected uniformly at random from a range $[-\eta/2, \eta/2]$, where η is a constant.

Figure 5.2a shows a simulation of this model on a one-dimensional ring. In this model aggregations form and move more slowly as their size increases.

Alignment: Individuals align by adopting the same direction as their neighbors. In one dimension, Czirok et al. (1999) use

$$s = G\left(\frac{1}{|R_i|}\sum_{j \in R_i} u_j(t)\right), \text{ where } G(u) = \begin{cases} (u+1)/2 & for \quad u > 0 \\ (u-1)/2 & for \quad u < 0 \end{cases}$$

and e as in the attraction model above. The function G ensures that velocities of individuals equilibrate around either -1 or 1. Figure 5.4 gives examples of simulations of this model for different numbers of individuals. As density increases collective motion emerges in the form of a single large group of individuals all going in the same direction.

In two dimensions, Vicsek et al. (1995) let $\underline{s} + \underline{e}$ be a unit vector with direction given by the average angle of the vectors plus some random term. Specifically,

$$\underline{s} + \underline{e} = \begin{pmatrix} \cos\left(\sum_{j \in R_i} \theta_j(t) + \varepsilon\right) \\ \sin\left(\sum_{j \in R_i} \theta_j(t) + \varepsilon\right) \end{pmatrix},$$

where the θ_j are the directions of i's neighbors and ε is chosen uniformly at random from a range $[-\eta/2, \eta/2]$. Unlike the two models above, in Vicsek's model $\alpha = 0$, but the individual i is always included in the in the set R_i of neighbors. Thus each individual includes itself as a neighbor when averaging velocities. Figure 5.7 gives snapshots of simulations of this model for different magnitudes of noise. Noise plays the opposite role of density: for higher noise motion is less ordered.

Repulsion, attraction, alignment, and blind angles: Couzin et al.'s (2002) model involves three zones of interaction: an inner zone of repulsion, an intermediate zone of orientation and an outer zone of

(Box 5.B continued on next page)

attraction (figure 5.8a). The individuals have a blind angle behind them within which they do not respond to individuals that
would otherwise be in their orientation or attraction zone. The
rule for repulsion is simply that individuals move directly away
from nearby individuals. The rules for attraction and alignment
are similar to those described for the two simple models, but are
only active in their respective zones. Figure 5.8 investigates a three-
dimensional version of this model for different sizes of orientation
zones. Provided there is a sufficiently large blind angle, the group
goes through a transition from swarm to milling torus to a highly
aligned group.

of the cockroaches. As in the SPP model, cockroaches within the cluster
move much less than those outside of it.

Jeanson et al. (2005) developed a parameterized model based on experiments on groups of two to four cockroaches. The principle underlying this model was similar to the simple aggregation SPP model, but
it included more detail of walking trajectories in different parts of the
two-dimensional arena, probabilities of individuals starting and stopping walking, and the effect of collisions from different directions such
as front and behind. The model showed that local contacts alone were
sufficient for the rapid aggregation observed in experiments (figure 5.2c).

Whether animals aggregate depends on their environmental context
(Krause 1994; Krause & Ruxton 2002). Larger groups provide dilution
from predator attack and individuals in smaller groups get a larger share
of food discoveries (chapter 2). Hoare et al. (2004) found killifish group
sizes were significantly smaller in the presence of food odor and larger
in the presence of an alarm odor. To explain the behavioral mechanisms
that produced these observations they used an SPP model of fish interactions, with terms for repulsion, attraction, and alignment. They showed
that the observed change in group size distribution could be explained
solely by a change in the size of the interaction zone. The distance at
which a fish is attracted to another fish decreases in the presence of food
and increases in the presence of a predator. This study provides a nice
link between mechanism and function: the regulation of group sizes to
perceived risk results directly from a change in interaction radius.

The mechanisms underlying spatial aggregation have been studied
for a range of species: from midges (Okubo & Chiang 1974) and bark
beetles (Deneubourg et al. 1990b) to primates (Hemelrijk 2000). More
than twenty years since its publication, the review by Okubo (1986) still

provides the best synthesis of mathematical and empirical aspects of aggregation.

Alignment

Attraction alone cannot explain the dynamics of most animal flocks. In particular, the aggregative clusters formed by between-individual attraction move slower as cluster size increases (figures 5.1a and 5.2a,c). These observations are in direct contrast to those of fish schools, locust swarms, and migratory birds that, while remaining a cohesive group, move rapidly in the same direction. Indeed, it is the rapid propagation of directional information that characterizes these groups, and poses the greatest challenge to our understanding of them (Couzin & Krause 2003). How is it that a bird flock or a fish school can apparently turn in unison such that all members almost simultaneously change direction?

It was the pioneering experimental work by Radakov (1973) that first showed how changes in direction can be rapidly propagated by local interactions alone. He used an artificial stimulus to frighten only a small part of a school of silverside fish. The fish nearest to the stimulus changed direction to face directly away from it. As these fish changed direction they stimulated others nearby, but further away from the artificial stimulus, to also change direction. A "wave of agitation" spread away from the artificial stimulus (figure 5.3). This propagation of directional information was much more rapid than the displacement of the fish. The fish nearest to the stimulus moved less than 5 cm in the same time it took every fish within 150 cm of the stimulus to change direction to face away from the stimulus. Changes in direction propagated at speeds of up to 11.8–15.1 meters per second over distances of between 30 and 300 cm.

While not directly inspired by Radakov's work, the transfer of directional information was the key ingredient in the self-propelled particle models of Vicsek et al. (1995). In fact, Vicsek's model has only two ingredients determining the direction particles move: alignment to nearby particles and noise (box 5.B). Figure 5.4a–c shows examples of these simulations in one dimension for different particle densities. A central prediction of Vicsek's model is that as the density of particles increases, a transition occurs from disordered movement to highly aligned collective motion (Czirok et al. 1999; Czirok et al. 1997; Vicsek et al. 1995). Figure 5.4d–f shows how the mean direction, or the degree of alignment, of particles changes through time in a one-dimensional version of the model from box 5.B for three different particle densities. At low densities, the alignment remains close to zero (figure 5.4a, d). At intermediate densities, all particles adopt a common direction for a period of time but this

109

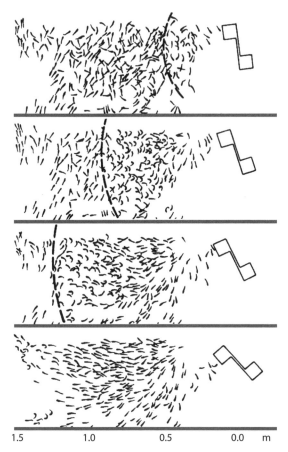

Figure 5.3. Example of Radakov's experiment where fish schools are presented with a fright stimulus. The position of fish was filmed and projected on a wall so that a picture could be made of the position and orientation of the fish. Reproduced from D. V. Radakov, 1973, *Schooling in the ecology of fish*, John Wiley & Sons.

direction switches at random intervals (figure 5.4b, e). At high densities, particles adopt a common direction, which persists for a long period of time (figure 5.4c, f). The transition from disorder (random motion) to order (aligned motion) occurs at a critical density, below which alignment is zero and above which absolute alignment increases with group size (Czirok et al. 1999).

Such a transition from disordered to ordered motion is seen in the collective motion of locusts. Buhl et al. (2006) looked at the alignment of various densities of locusts in an experimental ring-shaped arena. This setup effectively confined the locusts to one dimension and the degree

110

of alignment could be measured as the average direction of movement relative to the center of the arena. For small populations of locusts in the arena there was a low incidence of alignment among individuals. Where alignment did occur, it did so only after long initial periods of disordered motion (figure 5.5a). Intermediate-sized populations were characterized by long periods of collective rotational motion with rapid spontaneous changes in direction (figure 5.5b). At large arena populations, spontaneous changes in direction did not occur within the time scale of the observations, and the locusts quickly adopted a common and persistent direction (figure 5.5). As predicted by Vicsek's model, alignment of locusts becomes non-zero above a critical density (figure 5.6). The simplicity of Vicsek's SPP model suggests that phase transitions should be a universal feature of moving groups (Buhl et al. 2006). Similar transitions are observed in fish (Becco et al. 2006) and in tissue cells (Szabo et al. 2006).

When extended to two or three dimensions, Vicsek's model generates spectacular dynamical patterns that are highly reminiscent of the movement of flocks (figure 5.7). Again the two-dimensional model undergoes a phase transition where alignment becomes non-zero above a critical particle density or below a critical noise level (Vicsek et al. 1995).

While reproducing many of the characteristics of animal flocks, Vicsek's model is by no means sufficient to explain all aspects of flocking. To start with, it does not contain an attraction term of the type discussed in the previous section. In fish, attraction between individuals has long been viewed as having equal importance to alignment in determining group dynamics (Partridge 1982). The omission of attraction from Vicsek's model means that a bounded group cannot form. In an SPP model without an attraction term, a large group of particles moving in the same direction spreads out and particles will "escape" from the back of the group (Gregoire et al. 2003). When confined to a small space this diffusion will not lead to a significant breakup of the group because stragglers are picked up when they meet the large group again, but in an infinite (or large) space the group will eventually break apart.

A cohesive moving group can form if both attraction and alignment terms are included in an SPP model. Gregoire et al. (2003) drew a phase diagram for a two-dimensional SPP model that included terms for attraction, alignment, and noise. They found that when attraction was weak relative to alignment, particles behaved as either a disordered or moving "gas," similar to those seen in the two-dimensional Viscek model (figure 5.7). This gas was characterized by the proportion of particles that were members of the largest group being less than one. When attraction was increased, the proportion of particles within the largest group tended to one, and Gregoire et al. classified this state as a liquid "droplet." Within

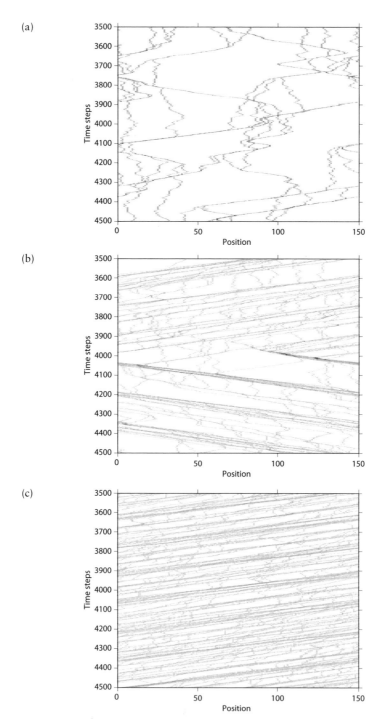

Figure 5.4. Example simulations from one-dimensional SPP models. Simulation of the SPP model of alignment in one dimension. The change in particle density through time for (a) $N = 10$, (b) $N = 50$, and (c) $N = 100$ particles. The alignment at time t is defined as

(d)

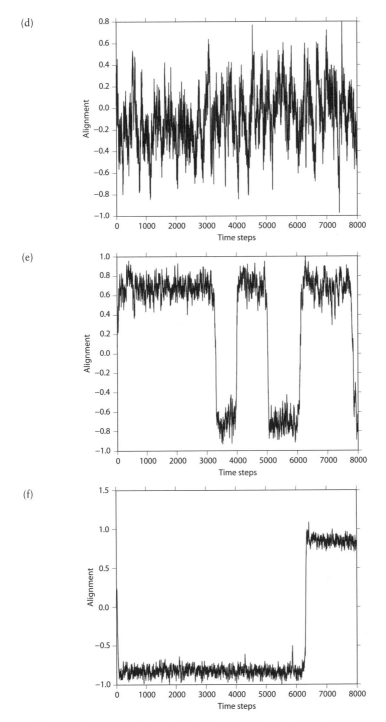

(e)

(f)

$1/n\sum_{i=1}^{n}\underline{u}_i(t)$, the average direction. The alignment is given for (d) $N = 10$, (e) $N = 50$, and (f) $N = 100$ particles. Other parameters are $L = 150$, $r = 1$, $v = 1$, $\alpha = 0.66$, and $\eta = 0.8$.

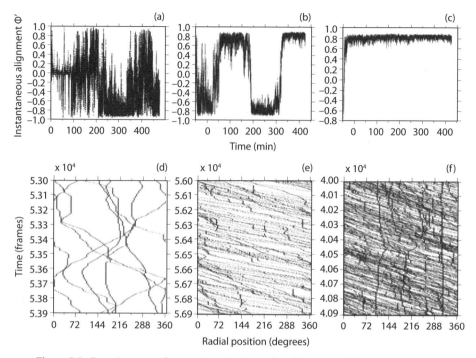

Figure 5.5. Experiments on locusts in a ring (reproduced from J. Buhl, D. J. T. Sumpter, I. D. Couzin, J. J. Hale, E. Despland, E. R. Miller, & S. J. Simpson, "From Disorder to Order in Marching Locusts," 2 June 2006, *Science* 312, 1402–1406, fig. 2, © The American Association for the Advancement of Science). The alignment over the experiment of (a) 7 locusts, (b) 20 locusts, and (c) 60 locusts. (d to f) Corresponding samples of time-space plots (3 min), where the x-axis represents the individuals' angular coordinates relative to the center of the arena, and the y-axis represents time.

this droplet two close together particles diffused away from each other through time while remaining within this large group. Compared to the gas in figure 5.7, in which groups split apart and reform, individuals moved around within the single droplet but did not leave it. As the attraction term was further increased, the liquid turned into a solid "crystal" and the particles remained at a fixed position within the crystal through time. Provided alignment was sufficiently large relative to noise, both liquids and solids exhibited cohesive collective motion where all particles moved as a group in the same direction.

A number of aspects of Gregoire et al.'s model resemble the motion of animal flocks. Moving crystals and droplets both exhibit periods of ballistic flight, where the mean square displacement of the group was proportional to $(\text{time})^2$, i.e., groups fly in a straight line. Furthermore, the lengths of these ballistic flights increased with the size of the group. This is

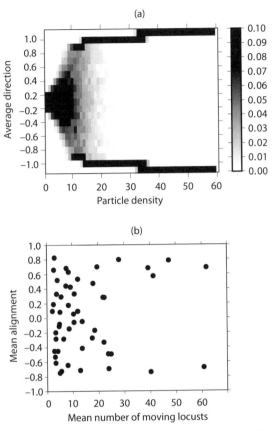

Figure 5.6. Comparison of the mean alignment in the (a) SPP model and (b) the locust data as a function of the number of particles (or locusts). Reproduced from Buhl et al. (2006).

in contrast to the non-moving phases where attraction is dominant, e.g., as in figure 5.2a. In this case, the mean square displacement of the group was proportional to time, and the lengths of ballistic flights decreased inversely proportionally to group size. Crystals and droplets both resemble various forms of moving animal groups: crystals look roughly like highly parallel groups of fish or birds, while the droplets possibly resemble flying locust swarms. Particularly interesting is the existence of mesoscopic "hydrodynamical" structures, such as jets, vortices, etc., within droplets (Gregoire et al. 2003). It is this dynamical patterning on a meso-scale within a generally coherent motion on the scale of the entire group that might be said to best characterize the collective motion of many flocking animals. However, the "zoology" of these meso-scale shapes has not been fully investigated and compared to empirical observations.

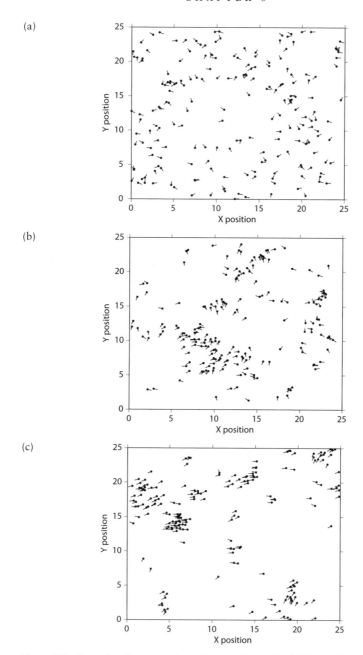

Figure 5.7. Example of patterns from the two-dimensional SPP model with alignment. "Heads" indicate position of the individual and "tails" give direction. Model is as described in box 5.B. Parameters are $n = 200$, $v_0 = 0.5$, $L = 25$, and $r = 1$. The noise is varied between simulations (a) $\eta = 3$, (b) $\eta = 1.5$, and (c) $\eta = 0.5$.

Rules of Motion

The attraction and alignment models discussed in the previous sections have not been calibrated against real data of how fish, birds, or locusts interact with one another. Instead, the philosophy of these models is to provide the simplest possible model that reproduces the key features of flocks. This philosophy is aimed at ensuring that model outcomes are not dependent on some particular biological feature, but reveal universal properties of all flocks. The approach is also to some degree unavoidable. Empirical determination of the detailed interactions of fish or birds is technically difficult. These groups move in two or three dimensions and often come in close contact with each other, making automated or even manual tracking difficult (Hale 2008).

There are, however, a number of high quality studies of fish interactions, most notable those of Partridge in the early 1980s. Studies of the structure of schools of saithe, cod, and herring show that fish maintain a minimum distance between each other, supporting evidence for local repulsion (Partridge et al. 1980). By tracking individual fish, Partridge (1981) established that saithe match their swimming direction and speed to their two nearest neighbors, but probably not to more distant neighbors. Partridge & Pitcher (1980) found that "blindfolded" saithe continued to match short-term changes in the velocity of their neighbors using their lateral line (the motion detecting sense organ that runs down fish bodies). Vision was, however, important in maintaining between-neighbor distance, with blind fish having increased nearest-neighbor distances. Fish that had their lateral line disabled compensated by changing postion so they could see direction changes by neighbors. In general, the lateral line appears to determine alignment, while vision determines attraction and repulsion.

An impressive step forward in the understanding of both the global structure of groups moving in three dimensions and the behavior of individuals within these groups is the Starflag project (Ballerini et al. 2008a; Cavagna et al. 2008a; Cavagna et al. 2008b). Using multiple cameras these researchers were able to determine the position of most of the starlings in flocks consisting of thousands of birds. Like fish, the starlings maintain a minimum distance from each other, i.e., have a zone of repulsion (Ballerini et al. 2008a). Starlings are also less likely to have neighbors behind or in front of them than to have neighbors on either side. As distance from a focal bird increases this spatial organization disappears, so that birds further away from a focal bird are equally likely to be at any angle.

Local spatial structure in starling flocks is not simply a function of distance but rather a function of neighbor number. The nearest neighbor is much more likely to be to the side of than directly in front of or behind a focal bird. This tendency then decreases for the second neighbor then the third neighbor and so on. After the sixth or seventh neighbor the spatial structure vanishes and these neighbors are equally likely to be at any angle relative to the focal bird (Ballerini et al. 2008b). This relationship is less robust when considering only the distance between neighbors. Even when the flock is more tightly packed spatial correlations are seen only between a fixed number of neighbors. The relationship would suggest that instead of interacting with all or some birds within a certain fixed radius, as is assumed in most models, starlings interact with their 6 or 7 nearest neighbors.

Complex Moving Patterns

The shapes of bird flocks, fish schools, and locust swarms are not limited to groups of aggregated or aligned individuals. Some of these shapes can emerge from simple interactions of repulsion, attraction, and alignment alone. For example, Couzin et al. (2002) proposed a model in which individual animals have three zones—repulsion, alignment, and attraction—of increasing size, so that individuals are attracted to neighbors over a larger range than they align, but decrease in priority, so that an individual always moves away from neighbors in the repulsion zone (figure 5.8a). These individuals also have a rear blind zone within which they cannot sense others.

Keeping the repulsion and attraction radii constant, Couzin et al. found that as the alignment radius increased, individuals would go from a loosely packed stationary swarm (figure 5.8b), to a torus where individuals circle around their center of mass (figure 5.8c) and finally, to a parallel group moving in a common direction (figure 5.8d). This transition from milling to torus to departure is typical of the motion of real fish schools. The model shows that these three very different collective patterns self-organize in response to small adjustments to one factor: the radius over which individuals align with each other.

Other patterns seen in animal flocks may be more difficult to produce from models of identical "memoryless" self-propelled particles interacting in a homogeneous environment. For example, Radakov (1973) reports "feeler" structures in silverside fish during their evening migration away from the shore. A few fish swim away from the group forming a ribbon-like structure as others follow. The leading group then reduces speed and starts feeding, at which point a "neck" builds up as more and more fish are drawn from the main group. In some cases this neck leads the whole group to the new feeding ground, while in others the neck breaks off and a sub-group separates from the main group. Overall, the

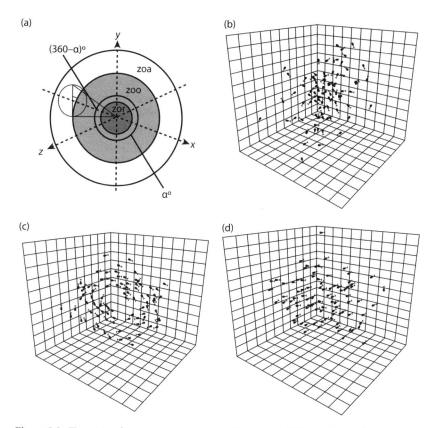

Figure 5.8. Transition from swarm to torus to alignment. (a) Illustration of the rules governing an individual in the fish model. The individual is centered at the origin: zor, zone of repulsion; zoo, zone of orientation; zoa, zone of attraction. The possible "blind volume" behind an individual is also shown as α, field of perception. Collective behaviors exhibited by the model: (b) swarm, (c) torus, and (d) dynamic parallel group.

process gives the impression of the school making a tentative investigation of whether it is worth moving feeding grounds.

Another common pattern in fish schools is the fountain response to the approach of a predator towards a group of prey (Hall et al. 1986). In this response the fish fan out in front of a predator and circle around behind it. Self-propelled particle models can reproduce this type of group response to predators (Inada 2002 and see Leading the Swarm, chapter 5). However, Hall et al. (1986) argue that a fountain response can occur simply by each individual prey moving away from the predator while keeping it at the edge of its field of view. Fish have a blind angle of roughly 60°, so by keeping the predator behind them at an angle of 150°

the fish are moving away from the predator as rapidly as possible without losing sight of it. This argument appears consistent with experimental data on the response of shoals of juvenile whiting (Hall et al. 1986), but it is not entirely clear whether social interactions may also play a role in creating the fountain effect.

Determining the degree to which simple rules for attraction and alignment capture the shapes produced by real animal groups remains a key problem (Parrish et al. 2002). No detailed statistical comparison has been made between the motion of and within real flocks and those predicted by SPP models. For example, Uvarov (1977) describes the marching bands of locusts as having a dense front and columns that go through an otherwise diffuse cloud of individuals. These observations have little in common with the shapes arising from, for example, Gregoire et al.'s (2003) model. Similarly, Ballerini et al.'s (2008a) observation that starling flocks have a dense boundary and a sparser interior directly contradicts most SPP models, which predict either homogeneous density within a group or a density that decreases with distance from the group's center. Explaining the emergence of complex moving structures will require greater consideration of the rules adopted by individuals, of how individuals interact with the environment, and of between-individual differences.

Decisions on the Move

When navigating, animals in moving groups usually have access to two types of information, their own experience or internal compass information and the direction taken by other group members. A central problem faced by animals traveling in these groups is how to integrate this information, especially when members cannot assess which individuals are best informed. In the context of avian navigation, two alternative schemes have been proposed (Wallraff 1978). The "many wrongs" hypothesis, which is described in more detail in chapter 4, is that individuals average their preferred direction, leading to a compromise in route choice. The average of these many wrongs should lead to an improvement in navigational performance. Wallraff's alternative to the many wrongs hypothesis is the "leadership" hypothesis. Under this hypothesis, one or a small number of the animals takes a leading role and the others follow.

Neither the many wrongs nor the leadership hypothesis accounts for how information is transferred between group members through local interactions. Indeed, the many wrongs hypothesis leads to the paradox, discussed in Integrating Many Wrongs, chapter 4, that for information to be transferred some individuals must follow others but at the same time too much following will reduce the success of the averaging. To bypass this limitation, Biro et al. (2006) developed a mechanistic model of navigational conflict between pairs of individuals. In the model (described in box 5.C), individuals interact according to two hypothesized forces: attraction to its own target position (own information) and attraction to the partner's current position (social information).

Figure 5.9a shows, for the model in box 5.C, the effect of varying the distance between the individuals' targets, d, on the final decision reached. The model predicts that at small distances between established routes, individuals average, with their position equilibrating at $d/2$. At a critical between-route distance, of approximately twice the range at which individuals are maximally attracted to their established routes, a bifurcation occurs. For d larger than this critical value, both individuals move closer to that of one of the individuals. A third possible outcome is splitting, where each individual moves exclusively towards its own target. Such outcomes occur over a wide range of d but always result from initial differences in the individuals' positions.

While the model in box 5.C provides an abstract representation of navigational decision-making, it was designed specifically with the behavior of homing pigeons in mind. Predisposition to a target models the phenomenon of route recapitulation and route loyalty by homing pigeons and between-individual attraction models social cohesion between birds. Biro et al. (2006) tested the model's predictions against data we collected on homing pigeons. Homing pigeons were first released repeatedly in order that they could establish their own route home from a release site. Once individuals had learned their own routes they were released in pairs. In these paired releases instances of many wrongs compromise and of leadership were observed, even within a single journey of a single pair of birds.

In order to test how the distance between the birds' "target" routes affected the outcome of their paired flight, the distance between the birds' independent flights were compared to the distance between their routes when in a pair. Figure 5.9b shows the largest and second largest modes of the distribution of distances between routes taken by individuals during their paired flight and the immediately preceding single (established) route as a function of distance between the birds' established routes at the corresponding point of the journey. We see a similar bifurcation in this data as we see in the model prediction (figure 5.9a). As the distance between the birds' targets increases a bifurcation occurs from compromise to leadership.

The model in box 5.C is limited because it deals with only two individuals and abstracts away possibly important aspects of spatial interactions. Couzin et al. (2005) proposed an SPP model where individual particles move in a two-dimensional space according to rules of attraction, alignment, and repulsion. In this model a large group of "uninformed"

Box 5.C Model of Paired Navigational Decision-making

We consider a dynamic model for decision-making, where two individuals, X and Y, each decide on a real-valued "position," starting from initial positions $x(0)$ and $y(0)$. These individuals come to a final position as a result of a combination of two forces: predisposition to move toward a target position and local attraction toward the other individual's current position.

Predisposition to target: X, respectively Y, is attracted to a target position with value 0, respectively d. The rate at which an individual moves toward its predisposed choice initially increases with distance from the target, but above a point of maximum attraction the rate decreases. For individual X, we model this rate with the function

$$-x\exp(-x/r_a),\qquad\qquad(5.C.1)$$

where x is the current position and r_a is the point at which the attractive force toward the target reaches a maximum. Individuals farther from the target than r_a have a weaker attraction toward it due to difficulties in perceiving the target, while individuals nearer than r_a have a decreasing but positive attractive force, modeling an increasing degree of "comfort" with decreasing distance to the target.

Between-individual attraction: We model this with the function

$$(x-y)\exp\left(-\left(\frac{(x-y)}{\sqrt{2}r_b}\right)^2\right),\qquad\qquad(5.C.2)$$

where x and y are the current positions of the two individuals and r_b is the point of maximum attraction to other individuals. Attraction only occurs locally, so that once individuals move out of the range of perception, the rate of attraction quickly decreases.

We combine the two forces acting on the individuals to give a differential equation model of how the individuals change position:

$$\frac{dx}{dt}=-x\exp(x/r_a)-\alpha(x-y)\exp\left(-\left(\frac{(x-y)}{\sqrt{2}r_b}\right)^2\right)$$

$$\frac{dy}{dt}=\beta(d-y)\exp(y/r_a)+\alpha(x-y)\exp\left(-\left(\frac{(x-y)}{\sqrt{2}r_b}\right)^2\right).\quad(5.C.3\text{ and }5.C.4)$$

The parameter α determines the ratio of the maximum between-individual attraction over the maximum attraction to the target. β determines the ratio $(Y\!:\!X)$ of the strength of the individuals' attraction to their targets. Figure 5.9a shows the equilibrium solutions to the model equations as a function of the distance d between the individuals' targets.

individuals interacts with two small groups of informed individuals that each move toward different targets. As the angle between the targets increases there is a bifurcation where the group goes from taking a direction intermediate to the two small leading groups to taking the direction preferred by one of the two groups.

Leading the Swarm

An interesting prediction of the Couzin et al. (2005) model is that a small number of informed individuals can lead a large group. In these simulations groups of 200 uninformed individuals were almost always successfully led to a target by groups of less than 10 leaders. Thus observations of large numbers of birds, fish, or insects moving in the same direction do not imply that even a majority of individuals know where they are going or even know which individuals know where they are going. The Couzin et al. (2005) model thus suggests a "subtle guide" mechanism: a largely uninformed group can be led by a small group of informed "leaders" even when the identity of the leaders is unknown.

One of the most impressive examples of a large group of uninformed individuals being led by a small group is the flight of honeybee swarms from their temporary bivouac on a tree branch to a new nest site (see Honeybee House-hunting, chapter 9). Up to around 10,000 bees of which only 2 or 3% are informed of the location of the nest site fly as a single swarm to the site. How does such a small group lead such a large group to a small nest site? Lindauer (1955) hypothesized that the informed individuals repeatedly "streak" through the swarm in order to inform the other bees of the direction of the nest. Janson et al. (2005) formalized this hypothesis in an SPP model and showed that 150 "streaker bees" could lead a swarm of 3,000 uninformed bees, and these swarms could avoid obstacles in their path without splitting. While streaking might help guide a swarm, the "subtle guide" hypothesis presented above suggests that streaking is not a requirement for a small number of individuals to lead a large swarm. A further alternative to the "subtle guide" or

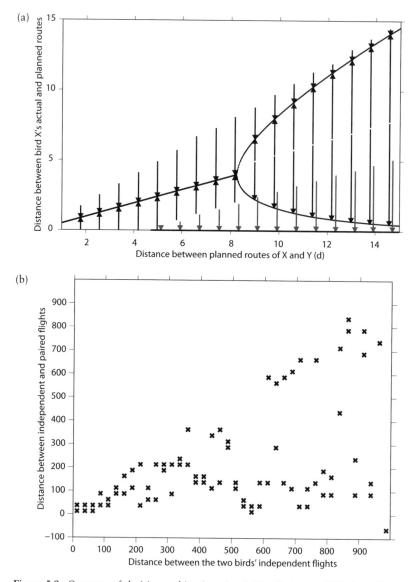

Figure 5.9. Outcome of decision-making in pairs. (a) Prediction model in box 5C. Equilibrium solutions of equations 5.C.3 and 5.C.4 as a function of the distance between the individuals' targets, d. The arrows show how different initial positions of bird X lead to different equilibriums. The initial position of bird Y is always $d/2$. The parameter values $r_a = 400$ and $r_b = 80$ were chosen to reflect the perception ranges of real pigeons. The other parameters $\alpha = 1$ and $\beta = 1$ assume no intrinsic difference between the birds. (b) Outcome of pigeon experiments. Histogram of point by point distances between each bird's established route and its taken route when in a pair, and plot of the largest and the second largest modes of the data.

"streaker bee" hypotheses is a "vapor trail," where the informed bees move to the front of the swarm and release a chemical pheromone creating a gradient that the other bees follow (Avitabile et al. 1975).

Beekman et al. (2006) tested the "vapor trail" hypothesis by sealing, in the bees, the glands that release pheromone and comparing the flight of sealed gland colonies with control colonies. Gland sealing had no significant effect on the flight speed of the swarm nor on the time it took the swarm to reach a nest box, contradicting hypotheses based on pheromones. Beekman et al. (2006) noted that some bees in the swarm were moving at maximum speed (9–10 m/s) while the swarm as a whole moved at only 2–3 m/s, providing evidence for the "streaker bee" hypothesis. Schultz et al. (2008) provided stronger evidence of streaking by filming a swarm from below. They found that bees in a top portion of the swarm flew quickly in the direction of the nest site and these fast moving bees were observed at the front, middle, and back of the swarm. However, while it appears clear that some bees streak along the top of the swarm and then return through it at slower speeds, there is still no direct link between these fast flying bees and the scouts.

Evolution of Flocking

Hamilton (1971) and Vine (1971) were the first researchers to look at how the geometry of an animal group might be shaped by natural selection. They both proposed "selfish herd" models in which individuals in the group are motivated to move into the center of the group by the risk of predation. In Hamilton's model, individuals live on a one-dimensional lattice and follow the rule: if the site an individual occupies has a larger population than sites to the left and right then it stays there, otherwise it moves to the neighboring site that is occupied by the largest number of other individuals. In contrast to the mechanistic model of aggregation described in box 5.A, Hamilton's model is motivated by functional considerations. However, the outcome of both models is similar: tightly packed clumps of individuals emerge (as they do in figure 5.2a). Vine and Hamilton both expand on this initial model and find similar results: tight aggregations are a consequence of selfish individuals' attempt to use other individuals as cover.

The geometrical predictions of selfish herd models hold for a wide range of species that form stationary groups (Krause 1994; Krause & Ruxton 2002; Quinn & Cresswell 2006; Rayor & Uetz 1990). Individuals near the center of these groups are less likely to be attacked than those on the edge. Several studies have revealed that when there is a predation risk, fish move closer together (Krause 1993; Tien et al. 2004). On the other hand, Focardi & Pecchioli (2005) found that the foraging success of deer increased with distance from the center of the group. There is thus a trade-off between increased food intake on the outside of the group and increased safety in the center. We might then expect position in a group to be determined by nutritional state, with well-fed individuals moving towards the center and hungry individuals moving to the outside.

In moving groups it is less clear how the position in a group relates to safety from predation. Parrish (1989) showed in laboratory experiments that grouping silverside fish are attacked less often by sea bass than stragglers that have recently departed from the group. However, in contrast to the idea that those fish in the center of the group can use those on the outside as a shield, Parrish also found that when the group is attacked it is the fish in the center that are targeted by the predators. Parrish suggested that this is because the predators attack the center of the group, which then splits in two leaving central individuals exposed. This interpretation is supported by simulations of SPP models (Inada & Kawachi 2002). Parrish's study is limited, however, by the fact that very few attacks by the predators were successful: only five group members were killed throughout all experiments, three of which were in the center and two on the periphery.

The complex dynamic patterns generated by flocking animals should convince us that a selfish desire to be shielded by others is not the only evolutionary force that has shaped them. Group membership may also allow individuals to gain information about the location of food (Pitcher et al. 1982) and of predators (Treherne & Foster 1981), to benefit in terms of energetic efficiency (Weimerskirch et al. 2001), and even to hunt co-operatively (Partridge et al. 1983). A problem, however, is disentangling functional and mechanistic explanations for dynamic patterns. Many patterns may be a consequence of the interactions between individuals and have little or no adaptive significance (Parrish et al. 2002). For example, the transition from disorder to order in locust marching appears to be a fundamental property of SPP models, suggesting that rather than resulting from the fine tuning of natural selection it is simply a necessary feature of all moving animal groups (Grunbaum 2006). Similarly, it would be wrong to conclude that a moving fish torus has evolved to signal among group members that departure is imminent, but rather it could be an unavoidable consequence of all members increasing their tendency to align with each other (Couzin & Krause 2003).

Behaviors that produce flocking patterns are in some cases themselves subject to natural selection. For example, one intrinsic property of SPP models is dynamic instability. Such instability was seen at intermediate densities in experiments on locusts, with changes in direction rapidly spreading through the entire group (figure 5.5e). If a small number of locusts spontaneously change direction, the others rapidly change their direction in response. This spread of directional information is reminiscent of Radakov's (1973) experiments on fish. Information about the presence of a stimulus is rapidly transmitted through the entire group.

Several modeling studies have investigated how the rules governing the alignment, repulsion, and attraction of self-propelled particles might be optimized so as to allow the particles to avoid predation (Inada & Kawachi 2002; Lee 2006; Lee et al. 2006; Zheng et al. 2005). In these studies a predator particle that is introduced into the simulation attempts to attack the group of prey particles. Inada & Kawachi (2002) varied the maximum number of neighboring individuals with which each prey aligned. They showed that if prey aligned with only one nearest neighbor then group movements were uncoordinated in response to a predator, but if they interacted with two or three the group was able to effectively align away from the predator. However, if prey individuals align with larger numbers of neighbors then the group would change direction slowly in response to a predator, because the minority of individuals that had sensed the predator and begun to move away from it would be "outvoted" by the uninformed majority that continue in their previous direction. Zheng et al. (2005) obtained similar results to Iwada by changing a different model parameter. They showed that there is an optimal weighting that individuals should put on aligning with other prey relative to orienting away from the predator. By aligning with each other rather than purely away from a predator, the prey avoid costly collisions. The collective outcome is a confusion effect, whereby the predator repeatedly changes target without successfully focusing on and killing one particular prey.

Most modeling studies of predator avoidance have looked at group success, measured in terms of number of group members captured by a predator, as a function of model parameters. From a functional viewpoint, however, the question is how individuals regulate their propensity to align, or their interaction range, or other aspects of their behavior so as to minimize their own probability of being caught by the predator. While aligning with others may increase the confusion effect for the predator, the best strategy for a focal individual may be to move directly away from the predator. As a result a social parasitism dilemma arises: while co-operating individuals can generate a pattern that optimizes group success, a defecting individual surrounded by co-operators can benefit to the greatest degree by not participating in the pattern. The pattern is then not evolutionarily stable (see chapter 10).

Wood et al. (2007) investigated the evolutionary stability of self-propelled particles to predation. They used the same model for particle movements as Couzin et al. (2002) but allowed the particles to evolve their interaction zones in response to predation. The main parameters

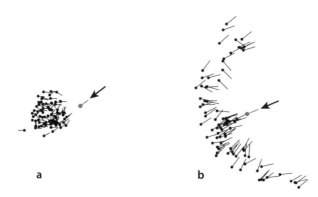

Figure 5.10. Typical example of the two types of evolutionarily stable flock types in the Wood and Ackland (2007) model. Each flock is shown before and during the attack of a predator; (a) is a compact milling torus that responds relatively slowly to the predator, while (b) is a dynamic parallel group with a high degree of alignment but only loose between-individual attraction. When a predator attacks, the group fans out to avoid it. Prey heads are marked with a circle and the line indicates their current velocity. Predators are larger in gray and marked with an arrow.

governing the interaction zones are the relative size of the attraction, R_a, and orientation zones, R_o, as well as the angle θ over which the particles can "see" their neighbors. The total area over which a particle could monitor its neighbors, i.e., $\theta \pi R_a^2$ was fixed to a constant for all particles. This constraint means that their viewing area is restricted to a local neighborhood of constant area. On the first generation a population of 80 individuals each with its own values of R_a, R_o, and θ was simulated for a sufficient number of time steps so as to allow a dynamic pattern to form. A predator, which attempted to capture the prey individuals, was then introduced into the simulation. After a fixed number of time steps those surviving individuals, i.e., those that had not been caught by the predator, went on to the next generation and those individuals that were caught were replaced by "offspring" of the surviving individuals. These offspring were subject to small mutations in the parameter values so that individuals with new values for R_a, R_o, and θ entered into the population.

There was a clear pattern in the evolution of the parameters. Firstly, the angle over which the particles could see evolved to be large, $\theta \approx 280°$ leaving a blind angle of 80°. This is reasonably close to the blind angle of 60° of many species of fish (Hall et al. 1986). The evolution of the small blind angle constrained the attraction radius R_a within which the orientation radius R_o was then free to evolve. Two evolutionary outcomes were possible for R_o, evolving either to be close to, but slightly larger than 0, or to be close to, but slightly smaller than R_a. In the first case the particles formed a slow moving milling group (figure 5.10a) while in the second they formed a fast moving dynamic group (figure 5.10b). Which of these outcomes evolves depends on the initial values of R_o within the population and the rate of mutation during selection. If R_o was initially large, a dynamic group would evolve and if it was initially small, a slow moving mill would evolve.

While both evolving through "natural selection," the dynamic group was more efficient than the slow moving mill at avoiding predation. The dynamic group had similar responses to predators as the optimized groups of Inada & Kawachi (2002) and of Zheng et al. (2005). It produced a confusion effect and split to avoid predation in 60–70% of cases. On the other hand, the predator was almost always successful in catching prey when faced with a slow moving mill. Wood et al.'s (2007) study is important because it provides evidence that complex collective level phenomena can evolve between "selfish" individuals without the need to invoke arguments based on kin selection or repeated interactions between individuals.

Synchronization

Synchronization occurs when large numbers of individuals co-ordinate to act in unison. In this wide definition of the word, many different types of collective behavior are examples of synchronization. A highly aligned group of birds, fish, or particles can be said to have synchronized their direction of movement. More commonly, however, when we use the word synchronization we are thinking about time. Bank robbers synchronize their watches before a robbery, the instruments of the orchestra are synchronized by the conductor and the sound is synchronized to the pictures in a film. It is this narrower sense of the word synchronization I use in this chapter. How and why do behaviors become synchronized in time?

Given that synchronization is a specific type of collective behavior it should come as no surprise that it shares many properties with systems looked at in earlier chapters of this book. In particular, and unlike with the bank robbers or the orchestra, synchronization can be achieved without a leader or centralized control. As with other types of collective behavior, we can also build mathematical models that describe how synchronization emerges from individual interactions. Indeed, some of the models of synchronization are among the most elegant models of collective behavior and have been employed successfully in understanding a wide variety of biological and social systems.

Rhythmic Synchronization

While the instruments of a concert orchestra are, at least in part, synchronized by signals from the conductor, the applause of the audience after the performance is not usually centrally controlled. Despite the lack of a central controller, in Eastern Europe and Scandinavia this applause is often rhythmical, with the entire audience clapping simultaneously and periodically. Neda et al. (2000a, 2000b) recorded and analyzed the clapping of theater and opera audiences in Romania and Hungary and found a common pattern: first an initial phase of incoherent but loud clapping,

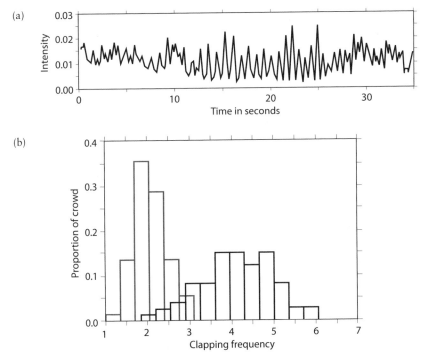

Figure 6.1. The emergence of synchronized clapping (reproduced from Neda et al. 2000a). (a) The average noise intensity of a crowd through time. The first 10 seconds show unsynchronized fast clapping, followed by a change to regular slower clapping, until around 27 seconds, followed by unsynchronized clapping again. (b) A normalized histogram of clapping frequencies for 73 high school students (isolated from each other) for Mode I (solid lines) and Mode II (zig-zag lines) clapping.

followed by a relatively sudden jump into synchronized clapping that, after about half a minute, was again rapidly replaced by unsynchronized applause (figure 6.1a). A surprising observation was that the average volume of the synchronized clapping is lower than that of unsynchronized applause, both before and after the synchronized bouts. While an audience presumably wants to maximize their volume and thus their appreciation of the performance, they are unable to combine louder volumes with synchronized clapping.

Neda and co-workers went on to record small local groups in the audience and asked individuals, isolated in a room, to clap as if (I) "at the end of a good performance" or (II) "during rhythmic applause." Both modes of clapping were rhythmical at the individual level, with individuals clapping in mode I twice as fast as those clapping in mode II (figure 6.1b). The important difference was in the between-individual variation for the two modes. When asked to clap rhythmically, isolated individuals chose similar, though not precisely identical, clapping frequencies, while when given the freedom to applaud spontaneously the chosen frequencies spread over a much wider range.

To interpret this observation, Neda et al. (2000b) used a classical mathematical result about coupled oscillators. Kuramoto (1975) studied a model of a large number of oscillators, each with its own frequency but coupled together so that they continually adjust their frequency to be nearer that of the average frequency. Kuramoto showed that provided the oscillators' initial frequencies are not too different, they will eventually adopt the same frequency and oscillate synchronously (Kuramoto 1984). This is what happens to audiences clapping according to mode II. Their initial independent clapping frequencies are close together, and by listening to the clapping of others, they synchronize their clapping. Audiences clapping in mode I have initial frequencies that are less similar to each other. Thus even if they try to adjust their clapping in reaction to the sound around them, the Kuramoto model predicts that they will never arrive at a state of synchronized clapping. This is exactly what happened in the recorded audiences: faster clapping, with greater inter-individual variation never synchronized. Concert audiences are thus forced to choose between two different manners of showing their appreciation: loud, frequent, unsynchronized or quieter, less frequent, synchronized clapping.

The importance of Kuramoto's model, which is presented in detail in box 6.A, is that it shows that individuals with slightly different frequencies can synchronize, each by moving their frequency slightly towards the average. It further predicts that above some critical level of between-individual variation synchronization does not occur at all (figure 6.2). In their empirical study, Neda et al. proposed that the opera crowds with unsynchronized clapping have a level of intrinsic variation above this critical level, and those with synchronized clapping have an intrinsic variation below the critical level. The switch in clapping mode from I to II reduces the between-individual variation and synchronization ensues.

Synchronized rhythmic activity is seen in many different animal groups and across much of biology (Strogatz 2003). As discussed in chapter 1, the oestrus cycles of female lions are usually synchronized within a pride (Bertram 1975) and the phase of these oscillations can be reset by the takeover of the pride by a new male (Packer & Pusey 1983). Likewise, human females' menstrual cycles become synchronized when the females are living or working closely together (Stern & McClintock 1998).

For many systems there is a good understanding of the physiological mechanisms involved in coupling individuals. Probably the best understood mechanism within animal behavior is the simultaneous flashing of some species of fireflies (Buck 1988; Buck & Buck 1976). Isolated fireflies

(a)

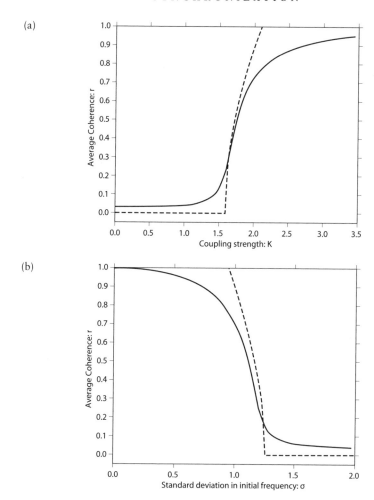

(b)

Figure 6.2. How coherence changes in the Kuramoto model with (a) coupling strength and (b) oscillator variation. The solid lines are the average coherence after 2000 time steps over 1000 runs of $N = 800$ oscillators. The dotted line in (a) is the approximation $r \approx \sqrt{\pi(K - K_C/K_C)}$, where $K_C = \sqrt{2^3/\pi\sigma}$. In this simulation $\sigma = 1$. (b) The dotted line is the approximation $r \approx \sqrt{\pi(\sigma_C - \sigma/\sigma)}$, where $\sigma_C = \sqrt{\pi/2^3}K$. In this simulation $K = 2$.

flash approximately once every second, but if they are subjected to an artificial flash at some point between flashes then their next flash is suppressed until approximately 1 second after the artificial flash (Buck et al. 1981). Mirollo & Strogatz (1990) developed a model to show that such phase resetting oscillators will synchronize, although they deal only with the case where all oscillators have the same intrinsic frequency. A number of good reviews have been written both of firefly flashing (Buck & Buck

133

Box 6.A The Kuramoto Model

Kuramoto (1975; 1984) proposed a simple model for the synchronization of coupled oscillators. Kuramoto assumed that the frequency, i.e., the rate of change of the phase θ_k, of each oscillator k, was determined by

$$\frac{d\theta_k}{dt} = \omega_k + \frac{K}{N} \sum_{j=1}^{N} \sin(\theta_j - \theta_k), \qquad (6.A.1)$$

where N is the number of oscillators and K is the strength of coupling between the oscillators. ω_k is the natural frequency of the oscillator, the frequency it will adopt if it is not coupled to other oscillators (i.e., when $K = 0$). Under this model, each oscillator adjusts its frequency in response to the phases of the other oscillators. If oscillator k has a smaller phase than the average phase of all other oscillators, then it will increase its frequency and thus become more in phase with the other oscillators. Likewise, if oscillator k has a larger phase than average, then it will decrease its frequency.

Intuitively, we would expect such a regulation to result in the phases of the oscillators becoming more similar. What is less clear is how we expect the degree of synchronization to change as a function of the coupling strength, or of the initial frequency differences between the oscillators. Kuramoto defined the coherence, i.e., the degree of phase synchronization, between the oscillators to be the complex number

$$re^{i\psi} = \frac{1}{N} \sum_{j=1}^{N} e^{i\theta_j}, \qquad (6.A.2)$$

where $i = \sqrt{-1}$. By recalling the definition of a complex number, $re^{i\psi} = r(\cos(\psi) + i\sin(\psi))$, and doing some algebraic manipulation we see that

$$\psi = \frac{1}{N} \sum_{j=1}^{N} \theta_j \text{ and } r = \frac{1}{N} \sqrt{\left(\sum_{j=1}^{N} \cos\theta_j \right)^2 + \left(\sum_{j=1}^{N} \sin\theta_j \right)^2}.$$

Thus ψ gives the average phase and r is a measure of the variation between the phases of the oscillators. When all the oscillators have the same phase, $\theta_j = \psi$, then

$$r = \frac{1}{N}\sqrt{N^2 \cos^2 \psi + N^2 \sin^2 \psi} = 1.$$

If all oscillators have a random phase, independent of that of the other oscillators, then as $N \to \infty$, $r \to 0$. Thus larger values of r indicate a more coherent population of oscillators.

Figure 6.2b shows how coherence changes with the variation of the initial frequency, i.e., the standard deviation of the distribution of the ω_k. When the standard deviation is small r is large. As the standard deviation increases r decreases. There is a critical level of between-oscillator variation, above which there is no coherence and below which coherence begins to emerge.

The definitions of r and ψ not only provide a convenient way of measuring coherence, but also allow an elegant mathematical analysis of Kuramoto's original model. A more detailed mathematical discussion of Kuramoto's model is provided in Strogatz (2000) . Here I summarize some of the main results presented by Strogatz. If we multiply both sides of equation 6.A.2 by $e^{-i\theta_k}$ and then equate the complex parts we get

$$r\sin(\psi - \theta_k) = \frac{1}{N}\sum_{j=1}^{N} \sin(\theta_j - \theta_k).$$

Substituting this into equation 6.A.1 we see that

$$\frac{d\theta_k}{dt} = \omega_k + rK\sin(\psi - \theta_k). \qquad (6.A.3)$$

When the coherence r is small or the coupling K is weak then the pull away from the natural frequency is small. Conversely, strongly coupled oscillators with high coherence have a strong pull away from the natural frequency.

Kuramoto assumed an infinite number of oscillators with initial frequencies ω_k taken from a distribution with a symmetrical probability density function with mean $\psi = 0$, e.g., the normal distribution. From equation 6.A.3 we see that at equilibrium

$$\omega_k = rK\sin(\theta_k).$$

(Box 6.A continued on next page)

Those oscillators with $|\omega_k| < rK$ approach this equilibrium, while those with $|\omega_k| > rK$ "drift" without arriving at the equilibrium. Kuramoto went on to derive a number of useful results. For example, if the ω_k are initially normally distributed with mean 0 and variance σ^2 then synchronization occurs, i.e., $r > 0$, whenever $K > \sqrt{2^3/\pi}\sigma$. Below the critical value $K_C = \sqrt{2^3/\pi}\sigma$ the oscillators all act independently of each other. For values of K slightly above this critical value the proportion of the synchronized oscillators is

$$r \approx \sqrt{\pi\left(\frac{K - K_C}{K_C}\right)}.$$

Figure 6.2a compares this approximation with an average outcome of 1000 simulations of 800 oscillators.

1976; Camazine et al. 2001) and rhythmic synchronization in general (Strogatz 2003; Strogatz & Stewart 1993).

Stochastic Synchronization

Anyone who has seen a flock of sheep or a group of hens pecking in a farmyard knows that domestically farmed animals commonly synchronize their behavior. Hens (Hughes 1971), pigs (Nielsen et al. 1996), and sheep (Rook & Penning 1991) are just some examples of animals that feed simultaneously. While simultaneous feeding may in part be accounted for by synchronized circadian rhythms and environmental cues, it can also be due to increased feeding in response to the feeding of others. For example, Barber (2001) found that laying hens were more motivated to feed in the presence of feeding companions.

Collins & Sumpter (2007) looked at how the number of feeding chickens at a particular point in a commercial chicken house influenced the rate at which other chickens began feeding nearby. Chicken houses are large homogeneous environments where a supply of food is provided along a feeding trough. The food constantly moves along this trough, ensuring that the supply is equal at all points of the feeder and that, in the absence of other birds, no part of the environment is consistently more attractive to the birds than any other. Figure 3.7 shows how the rate at which chickens join and leave a point at a feeding trough changed as a function of the number of birds already at that point. As the number of

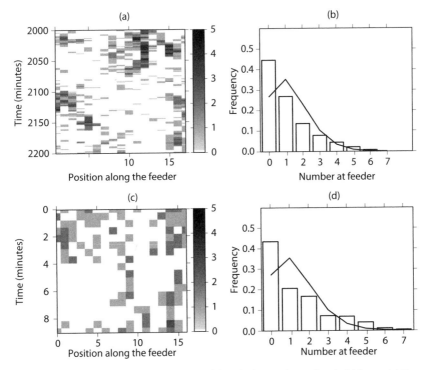

Figure 6.3. Comparison of simulation model and observations of real chickens. (a) Example of simulated number of birds feeding at different sections along the feeder through 200 simulated minutes. Darkness of shading indicates number of birds at that point along the feeder. (b) The distribution of number of chickens feeding per three adjacent feeding sections over 10,000 simulated minutes. (c) Example of activity at different sections along a real chicken feeder through 10 minutes. (d) Distribution of number of chickens feeding per three adjacent feeding sections over these 10 minutes. Fitted lines in (b) and (d) show distribution of the number of chickens assuming a Poisson distribution. See Collins & Sumpter (2007) for parameter values and details of the model and experimental setup.

chickens feeding at a point increases, the rate of arrival increases and the rate of leaving decreases.

Based on these observations, we developed a model to predict the long term dynamics of chickens arriving and leaving the feeder. The model predicted the dynamics of real chickens feeding (figure 6.3a) and the distribution of the number of chickens feeding at points along the feeder (figure 6.3b). Rather than being Poisson, as it would be if the chickens did not respond to the feeding of others, the distribution is skewed toward observations of either none or lots of chickens at the feeder. A qualitatively similar distribution was seen in further observations of the number of chickens at the feeder through space and time (figure 6.3c, d).

Box 6.B Stochastic Synchronization of Feeding

Box 3.C in chapter 3 describes a model of birds that choose to feed at a particular food patch as an increasing function of the number of birds already at that patch. Here, we consider a simulation of this model with access to only one food patch, $f = 1$. Birds can either forage at the food patch or rest away from the food patch. The probability of joining a food patch per bird not at the food patch is

$$\left(s + (m - s) \frac{C(1,t)^\alpha}{k^\alpha + C(1,t)^\alpha} \right), \tag{6.B.1}$$

and the probability per bird of leaving the food patch is a constant l (see box 3.C for an explanation of the parameters).

Figure 6.4b shows a time series of how many birds are visiting the feeder for a simulation of a group of $n = 7$ birds. Figure 6.4a shows the distribution through time of individuals at the food patch. There are bouts during which there are no birds at the food patch and bouts during which nearly all the birds are at the food. These resting and feeding bouts are relatively stable with most of the birds synchronizing their feeding.

To help understand how this synchronization arises, figure 6.4c shows the average number of birds joining the food patch per time step, i.e.,

$$\left(n - C(1,t) \right) \left(s + (m - s) \frac{C(1,t)^\alpha}{k^\alpha + C(1,t)^\alpha} \right) \tag{6.B.2}$$

as a function of the number of birds on the patch. The probability of one bird going to the feeder when none are there is relatively small, but once one bird is there, the probability is greater that another bird arrives than that the bird leaves. Furthermore, once two birds are there the probability of arriving becomes greater still until the number of birds climbs up to between 3 or 4. The average number of birds leaving the food patch per time step is $lC(1,t)$ as shown in figure 6.4c.

The points at which the average joining and leaving rates are equal correspond to feeding equilibriums. There are three such equilibriums, the smallest of which corresponds to a stable resting equilibrium and the largest to a stable feeding equilibrium. The middle equilibrium is unstable, such that if by chance the number of feeding birds drops below this unstable equilibrium then the birds quickly equilibrate at mostly resting. Alternatively, if the number of feeding birds goes above the unstable equilibrium, the number of feeding birds quickly equilibrates with 3 or 4 feeding. Thus the number of birds at the food patch jumps between the two stable equilibriums of none and, alternatively, 3 to 4 birds at the patch.

The sharp quorum response and positive feedback mean that, despite no environmental differences along different sections of the feeder, at any one time certain parts of the feeder will be preferred over others.

Box 6.B presents a simplified version of the chicken feeding model for a group of seven chickens visiting only a single feeder. Figure 6.4 illustrates how synchronized feeding occurs in this model. The birds alternate between most of them feeding and nearly all resting, but are not periodic in their feeding bouts. During resting periods there are small fluctuations with some individuals engaging in feeding. At some point these fluctuations take the number of individuals to a level at which the average rate of joining the feeder exceeds the average rate of leaving. The population quickly climbs to a point where the majority of birds are feeding. The number feeding fluctuates around this equilibrium, and at some point a large fluctuation leads to the number of foragers falling close to zero again. This pattern continues, but with no clearly defined frequency.

Gautrais et al. (2007) found that small groups of sheep synchronize their bouts of activity, and these bouts are not necessarily periodic. They fitted a Markov chain model to the data, where the state of the model was the number of active individuals. The measured transition probabilities of the Markov chain were such that the rate at which inactive individuals became active increased with the number of active sheep in the group and decreased with the number of inactive sheep. The opposite effect was seen on inactive sheep. These relationships produced rapid switching between the all-active and all-inactive states, without any obvious periodicity in the activity patterns.

From Randomness to Rhythm

The model in box 6.B is an example of synchronization without periodicity: there is no well-defined frequency in the activity patterns either of

139

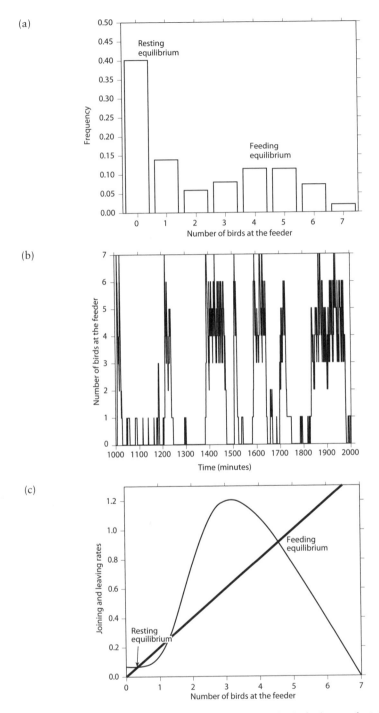

Figure 6.4. Outcome of model in box 6.B with $n = 7$ birds feeding at $f = 1$ food patch. (a) Distribution of number of birds feeding taken over 10,000 simulated time steps; (b) part of the time series of number of birds at the food patch; and (c) plot of the joining (equation 6.B.2) and leaving probability, $IC(1,t)$, as a function of the number of birds at the food patch, $C(1,t)$. Other model parameter values are, $s = 0.01$, $m = 0.4$, $k = 2.4$ and $\alpha = 4$.

individuals or of groups. Conversely, the Kuramoto model describes a situation where both individuals and groups have an inherent periodicity. Ants provide an interesting example of activity cycles that is not successfully modeled by either of these approaches. Figure 6.5 shows time series and frequency power spectra of both isolated individuals and whole colonies of the species *Temnothorax allardycei*. Single ants have no well-defined period between their bouts of activity, while whole colonies of these ants have synchronized, periodic activity bouts (Cole 1991a, 1991b).

A model based on box 6.B could be adopted to explain ant activity cycles. Under such a model ants would be assumed to have two states, active and inactive. We would further assume that encounters with active ants increase the probability of other active ants remaining active and/or the probability of inactive ants becoming active. Such a model certainly creates synchronization, but it is less clear how it generates periodicity.

How then can periodicity arise at the group level when absent in isolated individuals? An initial suggestion by Goss and Deneubourg (1988) was that after a bout of activity, inactive ants have an "unwakeable" period where interactions with others do not result in them becoming active. Under this model, isolated individuals' inactive bouts have a deterministic minimum time plus an exponentially distributed period until the start of the next active bout. In groups, this latter part of the inactivity bout can be interrupted by disturbance from other ants, whereby all ants are woken by the first ant coming out of its bout of inactivity. As a result, the inactive bouts become equal to the length of the "unwakeable" periods and the "unwakeable" periods become synchronized. An alternative assumption is that active ants are "unsleepable" with some minimum length of activity bouts (Cole & Cheshire 1996; Sole & Miramontes 1995; Sole et al. 1993). The effect of this assumption is similar to in the "unwakeable" model: although there was a minimum time between the start of activity bouts in isolated ants, the time between bouts has no well-defined period. When in groups, however, the ants became synchronized with the gaps between the starts of bouts determined by the "unsleepable" periods. These observations are largely consistent with observations of real ants (figure 6.5).

Activity bouts within ant colonies are not always periodic. Franks et al. (1990) found that colonies of *Leptothorax acervorum* have synchronized, but not always periodic bouts of activity. Boi et al. (1999) found that periodicity was strongest amongst workers further from the entrance to the nest and was disturbed by the return of workers to the colony. They also found that activity originated in the center of the nest, where the brood is kept, and spread outwards. In experiments where forager returns were prevented activity started first and lasted longer in the center of the nest than at the nest entrance.

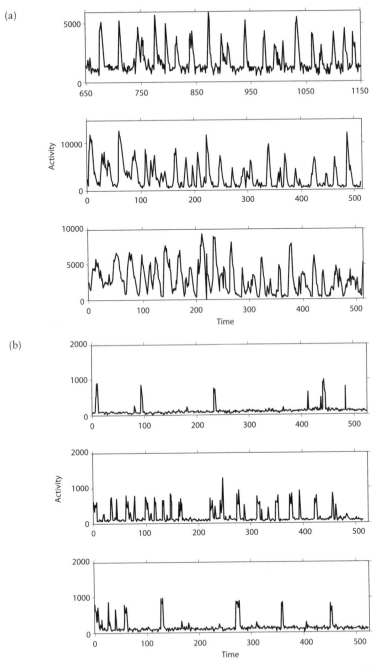

Figure 6.5. Activity records (taken from B. J. Cole, "Short-Term Activity Cycles in Ants: Generation of Periodicity by Worker Interaction," February 1991, *The American Naturalist* 137:2, 244–259, figs. 2 & 6, © The University of Chicago Press) for (a) three colonies of *Leptothorax acervorum* and (b) three isolated individual ants. One time unit is one minute.

Why Synchronize?

Synchronized activities are beneficial to individuals in a range of situations. For example, sheep (Ruckstuhl 1999), deer (Conradt 1998), and other ungulates have synchronized bouts of feeding and digestion. By choosing to forage for food together, individuals reduce their probability of being attacked by a predator and increase their opportunity for information transfer (see chapters 2 and 3). Synchronized activity could be a requirement for the social cohesion that allows animals to benefit from being in a group (Conradt & Roper 2000).

Box 6.C describes a simple functional model of a pair of animals, each of which chooses between either resting or foraging for food. The model assumes that the benefit of foraging decreases with the nutritional state of an individual. There is a cost associated with foraging, which is smaller if both individuals forage simultaneously. Two key points arise from the analysis of this model. The first point is that the evolutionarily stable strategy is for the individuals to synchronize their actions, i.e., forage at the same time and rest at the same time, even when their nutritional states are different (figure 6.7). The second point is that synchronization of actions does not imply synchronization of nutritional state. Instead, the individual with the lower nutritional state initiates foraging and the individual with higher nutritional state follows, because of the benefit it gains from foraging with a partner. Once the individual with higher nutritional state reaches a level of nutrition such that it no longer pays to forage, even with a partner, it will stop foraging. At this point the less well-nourished individual will also stop, because it does not pay to forage alone. The less well-nourished individual thus never "catches up" with the better nourished individual (figure 6.6b) and the two nutritional states remain at different levels.

The model has implications for how we think about leadership of groups. The less well-nourished individual in the pair takes the lead in initiating foraging, while the better-nourished takes leads in stopping foraging (Rands et al. 2003). These results are robust to the addition of noise. If the increase q and decrease r in nutritional state are random rather than constant on each time step then, on average, one of the individuals remains better nourished than the other (figure 6.6c). However, contrary to a suggestion by Rands et al. (2003), differences in nutritional states are not a consequence of synchronization. Rather, both individuals have a nutritional state that increases and decreases at the same rate: individuals with similar nutritional states maintain similar nutritional states on average, while those with different states maintain this difference on average. Any differences over time only arise through random drift. In other words, while nutritional state depends on how individuals

Box 6.C State-based Synchronization

Rands et al. (2003) proposed a model of a pair of foragers who, based on their own nutritional state and that of their partner, decide whether to forage for food or rest. Here, I present a simplified version of their model, which captures the essential features of the argument they present. Assume that on each day an individual must decide between foraging and resting. Foraging incurs a cost c due to predation risk but also gives a benefit b/s where b is a constant and s is the nutritional state of the individual. The benefit in foraging thus decreases inversely proportionately to nutritional state. Further assume that if an individual forages at the same time as its partner it gains benefit e from dilution of risk. Thus the payoff for foraging together with a partner is $b/s - c + e$ and the payoff for foraging alone is $b/s - c$. We assume that resting incurs neither benefit nor cost.

What strategy should an individual adopt? First consider the case where both individuals in the pair have the same nutritional state, s. In this case, if $s < b/c$ then it always is better to forage than to rest and both individuals will forage. If $s > b/(c - e)$ then it always is better to rest than to forage and both individuals will rest. For the intermediate values of $b/c < s < b/(c - e)$ the maximum payoff is obtained if both individuals forage. However, if the focal individual forages and its partner rests then the focal individual will get the lowest of the possible payoffs. Furthermore, if both individuals are resting then swapping to foraging without the certainty that your partner will also swap is costly. The individuals thus do best if they co-ordinate their foraging and resting, i.e., they synchronize their active and inactive periods.

When $b/c < s < b/(c - e)$, the evolutionary game defined in table 1 is known as a co-ordination game (see chapter 10 for more about evolutionary games). There are two evolutionarily stable strategies to such games and deciding which strategy will evolve is not straightforward. On the one hand it is optimal, in terms of higher benefits, for both individuals to forage. On the other hand, unlike resting, foraging is prone to errors resulting from one individual failing to co-ordinate. Here, I resolve this co-ordination issue by assuming that in repeated iterations over a number of days, whenever $b/c < s < b/(c - e)$ an individual will adopt the same behavior as it adopted on the previous day. Nutritional state can now be made time-dependent, i.e, represented by s_t. Individuals rest whenever $s > b/(c-e)$ or when $b/c < s < b/(c-e)$ and both individuals rested on the previous day, resulting in a decrease in nutritional state i.e., $s_{t+1} = s_t - r$. Individuals forage whenever $s < b/c$ or $b/c < s < b/(c-e)$ and the individuals foraged on the previous day, resulting in an increase in nutritional state, i.e., $s_{t+1} = s_t + q$. Figure 6.6a shows the outcome of such dynamics. The individuals' foraging bouts are periodic and synchronized and their nutritional states remain synchronized.

What happens if the individuals have different nutritional states, $s_{1,t}$ and $s_{2,t}$? Figure 6.7 summarizes the conditions under which it is optimal for a focal individual to co-operate given its own nu-tritional state and the state of its partner. If we assume that the individuals are aware of each other's nutritional state, then if $s_{1,t} > s_{2,t}$ both individuals will always forage when $s_{2,t} < b/c$ and $s_{1,t} < b/(c-e)$. Likewise, if $s_{1,t} > b/(c-e)$ and $s_{2,t} > b/c$ then both individuals rest. When $b/(c-e) > s_{1,t}, s_{2,t} > b/c$, it is best for individuals to co-ordinate. As in the case of identical nutritional states, however, it is not immediately clear upon which activity they should co-ordinate. Assuming as before that co-ordination is determined by the individual's previous action, the actions of the two individuals become synchronized (figure 6.6b). Interestingly, nutritional state does not become synchronized. Instead, foraging is initiated by the individual with the worst nutritional state, while resting is initiated by the individual with the best nutritional state. As a result, the individual with lower nutrition never "catches up" with its partner. If r and q are random variables, varying from day to day, instead of constant values the same pattern is seen (figure 6.6c). In general, there is no correlation between the nutritional states of the individuals despite a strong synchrony in their foraging patterns.

are synchronized, differences in nutritional state between the two individuals are largely independent of the degree of synchronization.

If each individual within a group has its own "ideal" point in time to perform an action, then as group size increases the heterogeneity of rhythms within the group also increases. Conradt and Roper (2003) used a model to show that in most situations it is beneficial to group members that decisions about the timing of events are made by consensus rather than "despotically." Their argument is based on the principle of

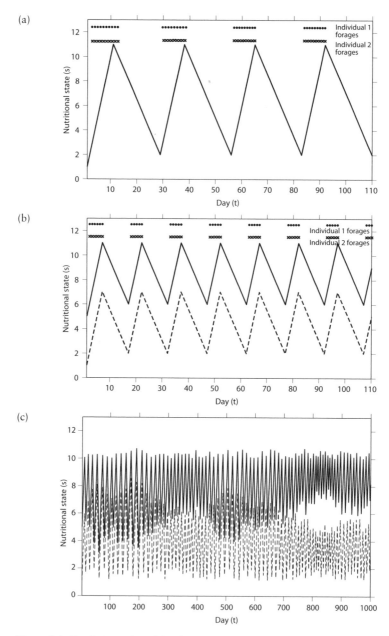

Figure 6.6. Simulations of model in box 6.C. Standard parameters are $b = 10$, $c = 5$, $e = 4$, $r = 0.5$, and $q = 1$. (a) Case where both individuals initially have the same nutritional state $s_{1,1} = s_{2,1} = 1$. The dots and crosses at the top of the figure indicate days on which individual 1 and respectively individual 2 foraged. (b) Case where both individuals have different nutritional states $s_{1,1} = 5$ and $s_{2,1} = 1$. (c) Simulations of the model where each time step q and r are selected uniformly at random with ranges [0.7, 1.3] and [0.2, 0.8] respectively (note different time scale).

146

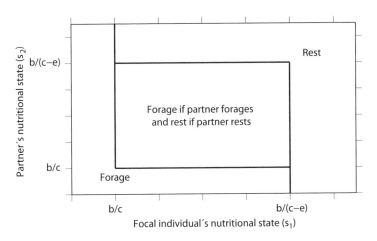

Figure 6.7. Evolutionarily stable strategy, according to the model in box 6.C, for a focal individual as a function of the focal individual's nutritional state and the partner's nutritional state.

"many wrongs" presented in box 4.B. It is better to time events according to the average preference, rather than adopting the preference of a single individual. Obtaining consensus among heterogeneous individuals is difficult, simply because different individuals want different things. Conradt and Roper (2007) develop an evolutionary game theory model of decision-making in groups of three or more members and show that over a wide range of conditions evolutionary stability of consensus can be obtained.

A prediction that arises both from Conradt and Roper's functional models and from Kuramoto's mechanistic model is that, if between-individual variation in the timing of events becomes too large then synchrony will break down. Conradt (1998) observed that male-only and female-only groups of red deer had more synchronized bouts of activity than mixed sex groups. This loss of synchrony could result from differences between males and females in the amount of food they need and the digestion time for this food (Ruckstuhl & Neuhaus 2002). Interestingly, within the mixed sex groups the female-female and male-male synchrony was much lower than that in the single sex groups. Similar observations have been made of alpine ibex (Ruckstuhl & Neuhaus 2001).

In terms of a functional explanation, desynchronization in mixed sex groups could be attributed to male harassment of females, competition between males in the presence of females, or some other social conflict within the group (Conradt 1998). However, such desynchronization is also explainable purely in terms of mechanisms. If the distribution of initial frequencies are bimodal, Kuramoto's model predicts that those individuals with initial frequencies nearer to the mean initial frequency will synchronize, while those with initial frequencies further away from the mean will remain close to their initial frequency (Strogatz 2000). Thus a proportion of the males become more synchronized with the females, increasing the degree of male-female synchrony, above that of separate groups, while decreasing male-male synchrony.

In Conradt's observations there were more females than males within mixed groups (Larissa Conradt, personal communication). Under these circumstances, Kuramoto's model further predicts that the pull on the males to synchronize with the females is stronger than the pull of the males on the females. This pull in opposite directions leads to desynchronization of both sexes, but greater desynchronization among males than among females. These predictions are confirmed in Conradt's observations of mixed groups, with male-male synchrony near to zero, female-female synchrony dropping slightly compared to single sex groups and male-female synchrony below that of female-female synchrony. Kuramoto's model provides us with null hypotheses about whether groups of diverse individuals will synchronize. It is only when these null hypotheses fail that we need to invoke additional functional explanations.

Temporal synchronization can produce patterns in the spatial organization of animal groups. Conradt (1998) hypothesized that the lack of synchrony in mixed sex groups leads to segregation of males and females. If females benefit from leaving an area before males then mixed sex groups are more likely to split than same sex groups. Ruckstuhl & Neuhaus (2002) found that sexual segregation was more common in ungulate species in which males and females were of different sizes and as a result had different activity budgets. The question of whether failure to synchronize is the primary explanation of spatial segregation in ungulates remains controversial. A number of studies have provided mixed results about the importance of synchrony in this context and suggest that spatial segregation is caused by a range of different factors (Calhim et al. 2006; Kamler et al. 2007; Loe et al. 2006; MacFarlane 2006). While temporal synchronization is unlikely to be a universal explanation of segregation, it remains an important factor in determining the spatial patterns produced by animal groups.

Anti-phase Synchronization

Kuramoto's model explains synchronization through locking of phases. Oscillators can, however, synchronize without adopting the same phase. Indeed, the first recorded observations of synchronization between pendulum clocks, made by their inventor Christiaan Huygens, were of out of phase synchronization (Bennett et al. 2002; Strogatz 2003). Huygens noticed that two pendulums hanging from the same beam became synchronized, such that when one pendulum was at its right extreme, the other was at its left extreme. This anti-phase synchronization occurs because of a weak coupling through lateral motion of the structure upon which the pendulums are mounted. Bennett et al. (2002) constructed an experimental setup with two pendulums, which allowed them to reproduce Huygens' findings. They developed a mathematical model to show that anti-phase synchronization is the only stable outcome for this system provided there is sufficiently strong coupling between the pendulums. As in the Kuramoto model, synchronization occurs even when the natural frequencies of the pendulums are slightly different.

An example of pairs or small groups of animals becoming anti-phase synchronized is sentinel behavior. McGowan & Woolfenden (1989) observed small groups of Florida scrub jays and noted when each member engaged in vigilance, looking around for potential threats, and foraging, looking for food. They found that vigilance was co-ordinated, such that periods of vigilance overlapped less than expected than if the decision to become vigilant was independent of the behavior of other individuals. One of the birds acted as a sentinel while the others fed. The periods of sentinel behavior were out of phase with each other, reducing the probability that a predator could attack unnoticed.

A functional explanation of sentinel behavior poses a challenge, because the individual keeping watch is losing the opportunity to forage for food. What is to stop the sentinel from cheating and skipping its turn to continue foraging instead? The dilemma here is similar to that of producers and scroungers (box 3.B) and is a typical example of social parasitism (see chapter 10). Although the group would have the least risk of predation were individuals to take turns being sentinels, for each individual the incentive is to often skip their turn to keep watch. Bednekoff (1997) proposed a simple solution to this problem based on selfish sentinels. He considered how nutritional state should influence the relative costs and benefits to an individual of foraging and sentinel behavior. Individuals that have just fed have a lower need for food and thus a greater incentive to take a safe position where they can keep watch. Bednekoff's model predicted that if being sentinel provided extra safety compared to foraging and that if a sentinel detected a predator this information spread to foragers then co-ordinated sentinels was an evolutionarily stable behavior.

Bednekoff & Woolfenden performed experiments on pairs of Florida scrub jays to test the plausibility of Bednekoff's model. They found that when scrub jays were fed they spent less time feeding and more time acting as sentinels (Bednekoff & Woolfenden 2003). Furthermore, the partners of the fed individuals reduced their time acting as sentinels and foraged more (Bednekoff & Woolfenden 2006). In pairs of scrub jays these responses can lead to a "seesaw" synchronization of feeding and sentinel behavior. Individual A finds food, eats it, and then begins to act as a sentinel, while individual B searches for food. When individual B has found and eaten food, its tendency to become a sentinel increases, while individual A's tendency to feed increases. In this case, and unlike the model of in-phase synchronization in box 6.C, anti-phase synchronization arises from and sustains differences in nutritional state.

Animals in larger groups also exhibit turn taking in sentinel behavior. Meerkats forage by digging into the ground, in a manner that makes it impossible to observe what is going on around them. The meerkats usually forage in groups. When not digging, some individuals stand guard, looking around for potential predators. Clutton-Brock et al. (1999) observed that meerkats take turns in guarding. When there was no guard then the probability that an individual would start guarding was twice as high as when there was a guard, and if two or more individuals happened to be guarding at the same time one of them would usually stop guarding relatively quickly. Turn taking was not in a consistent order, although individuals did not tend to take consecutive bouts of guarding. Like the scrub jays, an important factor in whether a meerkat would guard was whether it had been fed or not. Clutton-Brock et al. (1999) found that if they fed meerkats they would guard more often and forage less.

Clutton-Brock et al.'s (1999) experiments "provide no indication that the alternation of raised guarding depends on social processes more complex than the independent optimization of activity by individuals, subject to nutritional status and the presence (or absence) of an existing guard." Bednekoff's (1997) model provides an elegant mathematical demonstration of how this guard alternation can evolve. It also provides an extension of the seesaw concept and double pendulum anti-phase synchronization to groups of more than two individuals. Simply by aiming to maximize their own survival individuals will move out of nutritional phase with each other, so that there is usually a single guard with a high nutritional level.

— Chapter 7 —

Structures

In Småland in southern Sweden the ground is full of stones, carried down during the ice age. When the people of Småland began to farm the land they picked up the stones in an area, lay them in piles at the side and planted their crops. As the years went by they ploughed the fields and more stones came up and the piles turned into walls. As the walls expanded they joined each other, separating the land into a patchwork of separate fields. If you go to Småland today, you will find well-ploughed fields surrounded by thick walls with occasional piles of stones near the middle of the fields. Although some of the walls built in Småland must have involved planning, such a plan was not a requirement for their construction.

This story illustrates a central idea of this chapter: co-ordinated construction can be achieved by a group of workers without direct communication among them. The patchwork of fields separated by stone walls emerged from local interactions between the farmers and their environment. In this chapter we look at a number of spatial structures that can arise through local interactions between animals and their environment. Pillars and chambers in the nest of social insect colonies; rail, road, and supply networks in humans; and the pheromone trail systems of ants are all constructed in this way. Often the complexity of these patterns gives an impression of centralized design or planning. However, positive feedback operating in space can produce a rich variety of patterns without central organization.

Pillars and Walls

Theraulaz et al. (2003) propose that local interactions between animals and their environment, a mechanism they call *stigmergy*, combined with interactions with large scale environmental features, which they call *templates*, provide two of the key mechanisms for nest construction by insect societies. Examples of templates include light, temperature, and humidity gradients that determine the point at which an individual picks up or drops building material. Templates may also be determined by a signal from individual insects. For example, the queen of the termite *Macrotermes subhyalinus* emits a diffusive pheromone, which decreases in concentration with distance from the queen (Bonabeau et al. 1998b).

To illustrate the theory of how stigmergy and templates interact to produce collective structures Theraulaz has, together with various co-workers, investigated the construction of cemeteries by *Messor sancta* ants (Jost et al. 2007; Theraulaz et al. 2002, 2003). In experiments, a ring of ant corpses was placed around the edge of a circular arena with a hole in the center, through which living ants could access the arena. These worker ants tend to pick up corpses when they find them and then deposit them again at places where there are other corpses. As a result, clusters of dead ants build up through time.

Box 7.A presents a model based on the observation that ants are more likely to pick up and less likely to drop corpses where their density is low. Further assuming that corpse carrying ants perform a random walk, the model predicts that the emerging piles of corpses should be regularly spaced (figure 7.1). The pattern is a result of local activation and long range inhibition (Gierer & Meinhard 1972; Murray 1993). The local activation is a tendency of ants to pick up corpses in areas where local corpse density is low and move them to areas where corpse density is high. As a result, the construction of new piles near to those already established is inhibited. The worker ants move quickly resulting in a long range inhibition, and the distance between the piles eventually far exceeds the local range at which the ants can sense corpses. This inhibition decreases with distance from the pile. At some particular distance from an established pile the inhibition becomes less than the intrinsic probability of dropping a corpse, and another pile is established. As a result, there is a regular spacing between piles, the length of which depends upon the speed at which the ants move and the strength of inhibition. These predictions were confirmed in the experiments: the ants built piles of corpses around the arena wall at roughly regular intervals and the intervals between the piles increased in proportion to arena size (Theraulaz et al. 2002).

The addition of an environmental template can change the shape of the corpse piles. Jost et al. (2007) compared the behavior of individual ants in the presence and absence of air currents. The presence of wind increased the probability of ants picking up and decreased the probability of them dropping corpses. The experimenters also showed that when an air current is induced across a circular pile of ant corpses the wind strength is lowest behind and in front of the pile (the front of a pile being the side which faces towards the wind), but highest at the sides of the pile. Thus ants are more likely to remove corpses from the sides of the

Box 7.A Local Activation, Long-range Inhibition

How systems of interacting units generate robust spatial patterns is a central question at all levels of biology. One of the most revolutionary ideas in the theoretical study of pattern formation was Turing's recognition that passive diffusion of two or more chemicals combined with reactions between the chemicals to produce spatial patterns, even if the chemical reaction has no pattern forming qualities in the absence of diffusion (Turing 1952; reviewed in, for example Britton 2005 or Murray 1993). Up until the publication of his 1952 paper, diffusion was usually thought of as a mechanism for the dispersal rather than the generation of patterns. Gierer and Meinhardt (1972) went on to show that non-linear reactions involving two chemicals, one activating and the other inhibiting, would spontaneously produce patterns. The key condition for pattern formation in these activator-inhibitor systems is that the activator chemical diffuses slowly and has a positive effect on its own growth as well as that of the inhibitor, while the inhibitor diffuses more quickly and inhibits both the activator and itself. As a result, small regularly spaced spikes of activator chemical accumulate, which through the local generation of fast diffusing inhibitor have a long-range inhibitory effect on the production of further spikes nearby.

In the case of ant cemeteries, local activation is mediated through the ants' decision whether to pick up or drop an ant corpse at a particular point. Theraulaz et al. (2002) proposed that the rate at which corpses would be deposited at a particular point x should be proportional to

$$f(c(x), a(x)) = a(x)\left(k + \frac{\alpha_1 C(x)}{\alpha_2 + C(x)}\right) - \rho\frac{\alpha_3 c(x)}{\alpha_4 + C(x)}, \quad (7.A.1)$$

where $a(x)$, ρ, and $c(x)$ are respectively the density of carrying ants, non-carrying ants, and corpses; $\alpha_1, \alpha_2, \alpha_3, \alpha_4$, and k are constants and

$$C(x) = \frac{1}{2\Delta} \int_{x-\Delta}^{x+\Delta} c(z)dz$$

is the average number of corpses within a local region of the point x. Δ gives the size of the range over which active ants can detect

(Box 7.A continued on next page)

corpses. Experimental observations determined that this distance was small (0.5 to 1 cm) compared to the perimeter of the experimental arena (25 or 50 cm). The first term in equation 7.A.1 models an increase in the rate of dropping corpses with increasing local corpse density. The second term models a decrease in the rate at which non-carrying ants pick up corpses with local density of corpses. This second term requires some extra consideration since it assumes that the density of non-carrying ants, ρ, is constant over the whole arena. This assumption is made despite the fact that whenever a carrying ant picks up a corpse the number of available non-carrying ants should be depleted. In the experiment this assumption may be justifiable by the continual flux of ants in and out of the arena from a central hole.

Given equation 7.A.1 for corpse dependent activation, the following reaction-diffusion model expresses the rate of change of corpse and active ant density along the arena's perimeter

$$\frac{\partial c}{\partial t} = f(c,a)$$

$$\frac{\partial a}{\partial t} = -f(c,a) + D\frac{\partial^2 a}{\partial x^2}. \qquad \text{(7.A.2 and 7.A.3)}$$

The second term in equation 7.A.3 expresses the assumption that, when carrying the corpses, ants perform a random walk with diffusion co-efficient D. These equations fulfill the requirements of an activator-inhibitor system. The presence of corpses increases the rate at which corpses are deposited by carrying ants. The carrying ants diffuse faster than the corpses and inhibit the growth of further corpse piles.

Figure 7.1 shows numerical integrations of equations 7.A.2 and 7.A.3 for different densities of corpses and arena sizes. In these simulations, $\Delta = 1$ cm, implying only local activation. The resulting equilibrium distance between the corpse piles is between 15 and 25 cm, depending on the density of corpses. Corpse piles thus inhibit the construction of further piles over a long range, despite the local interaction of the ants with the corpses. The model further predicts that the number of piles increases with arena size and with the density of corpses. It was these predictions that were tested and confirmed in the Theraulaz et al. (2002) experiment.

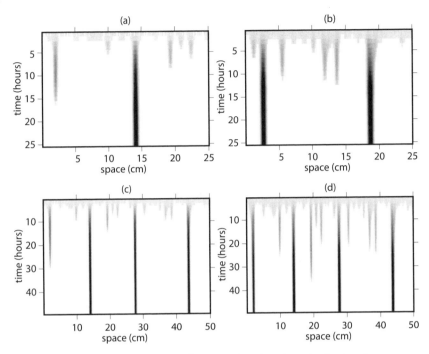

Figure 7.1. Simulation of the model in box 7.A. Plots show how the density of ant corpses around the perimeter of the arena changes through time. Initially corpses are distributed uniformly at random around the edge of the arena (denoted by light grey). As the active ants pick up and move corpses, they gather first into many (dark grey) then a small number (black) of clusters. The number of corpse piles depends on the number of corpses and the arena size: (a) 100 corpses in an arena of perimeter 25 cm; (b) 200 corpses/25 cm arena; (c) 200 corpses/50 cm arena; and (d) 400 corpses/50 cm arena. (Simulation and figure created by Stam Nicolis; for details of other parameter values see Theraulaz et al. 2002.)

pile. From these observations of individual ant behavior and of air flow, Jost et al. predicted that elongated walls would be built in the direction of the air current. This is exactly what happened in experiments with large number of ants (figure 7.2). Pillars became regularly spaced walls when the ants built in the presence of an air current.

Neither stigmergy nor templates are a requirement for the production of collective patterns. Symmetrical structures, such as the domes built by wood ants or the craters built by *Messor barbarus* ants can result from the independent actions of the colony's ants (Chretien 1996; Theraulaz et al. 2003). For example, Chretien (1996) showed that when an individual *M. barbarus* ant leaves the nest hole with a sand pellet, she moves in a straight line away from the hole in a random direction. Once the ant is, on average, 4.8 cm from the hole she drops the pellet. The fact that the

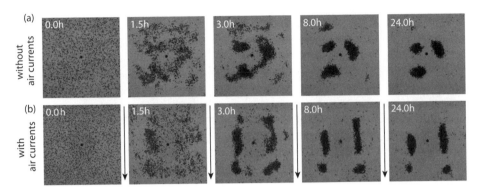

Figure 7.2. Spatio-temporal dynamics of corpse clustering (a) in the absence of and (b) in the presence of air currents. Black dots are ant corpses and black arrows indicate the air flow direction. (Reproduced from C. Jost, J. Verret, E. Casellas, J. Gautrais, M. Challet, J. Lluc, S. Blanco, M. J. Clifton, & G. Theraulaz, 2007, "The interplay between a self-organized process and an environmental template: Corpse clustering under the influence of air currents in ants," *Journal of the Royal Society Interface* 4, 107–116, fig. 4).

direction chosen by the ant is independent of the direction taken by the other ants in the colony produces a symmetrical crater (figure 7.3).

The even crater wall can be seen as a consequence of the central limit theorem discussed in chapter 4, box 4.B. Each individual ant makes an independent decision about which direction to take. As a result, the height of the wall at any point around the crater increases in proportion to the number of individuals. Furthermore, the standard deviation in the wall height increases as its square root. Thus, once the wall is reasonably high the variation in its height will be small relative to the average height around the top of the crater. Despite, and indeed because of, the ants working independently, an even outer wall is constructed.

Leptothorax ants also build circular nest walls (Franks et al. 1992). Franks and Deneubourg were the first to successfully provide a combined modeling and experimental approach to understanding nest construction. They showed that the walls originate from a combination of each individual's tendency to drop building grains at a fixed distance from the center of the nest with a stigmergic interaction: ants are more likely to leave grains where others have already been deposited (Franks & Deneubourg 1997). They further found that doubling, or in some cases, nearly tripling the number of workers in an established nest did not usually lead to an increase in the size of the nest, despite the fact that a nest built from scratch is likely to have an area proportional to the colony's size (Franks & Deneubourg 1997; Franks et al. 1992). Similarly, Aleksiev et al. (2007) manipulated the nest area of fixed size colonies and found that substantial

Figure 7.3. An example of a symmetrical crater wall built at the nest entrance to a colony of *Messor barbarus* ants (photo: Guy Theraulaz).

rebuilding only occurred when a nest was reduced to one-quarter but not to one-half of its standard size. This historical dependency can be accounted for by the reduced propensity to take building material from already established walls. It is only when the density of ants crosses a threshold that the walls are moved further from the nest's center.

Tunnels and Tents: Why Co-operate in Building?

Social insects' nests are in many cases built by sterile workers. These nests allow a reproductive queen to produce offspring that in turn provide indirect reproductive benefits to the workers. Collective construction is, however, not limited to sterile workers of social insects. For example, many bark beetles perform a simple form of co-operative tunneling in their attack on trees. Individual beetles arriving on a tree release a pheromone that attracts other bark beetles. Provided there are sufficient beetles in the vicinity of the pheromone release, this point becomes the focus of boring into the tree's bark by large numbers. Further pheromone releases induce a positive feedback loop and together the beetles can overcome

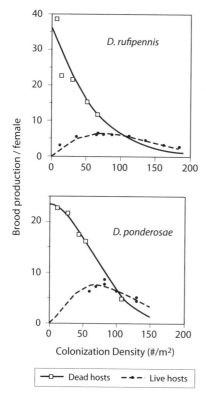

Figure 7.4. Reproductive success of bark beetles colonizing living vs. previously dead hosts for two species of bark beetle: (a) *Dendroctonus rufipennis*, and (b) *Dendroctonus ponderosae*. (Reproduced from K. F. Raffa, "Mixed messages across multiple trophic levels: The ecology of bark beetle chemical communication systems," April 2001, *Chemoecology*, 11:49–65, fig. 3, © Springer-Verlag). See Raffa (2001) for details of data sources and data fitting.

the tree's defences (Berryman 1999; Costa 2006). In large enough numbers, the beetles kill part or all of the tree, allowing them to lay eggs within the bark (Wertheim et al. 2005). The hatching larvae are then able to feed on inner bark tissue and mature to adulthood.

The fact that trees have extensive defense mechanisms against boring can help explain the evolution of co-operative digging. Although it can be costly to release pheromone, both in terms of its production and in the possibility of its attracting predators, a single beetle is unable to overcome the tree's defenses. Thus the costs of pheromone production are outweighed by the benefits of attracting others. Raffa (2001) reviewed the relationship between density and female brood production in 19 separate studies of bark beetle colonization. He found that in dead host trees per capita brood production decreased with density, while in living host trees per capita success initially increased with density and only decreased when densities became high (figure 7.4). The synergistic benefits of attacking the tree at the same point thus initially outweigh any costs of attracting others (Synergism, chapter 10).

Increasing reproductive success with group size may account for co-operative construction in other species of insects. For example, Austra-lian *Dunatothrips* create tent-like structures on the surface of leaves. Bono & Crespi (2006, 2008) compared survival and reproduction of thrips that founded structures alone with those in groups of two or more individuals. They found that although per capita brood production fell with group size, foundresses were more likely to survive to reproduce in groups than when alone. Several studies of other species of insects have concluded that foundress associations are beneficial to all parties (Bernasconi et al. 2000; Jerome et al. 1998; Tibbetts & Reeve 2003). It is likely that the relative success of groups is at least in part accounted for by a reduction of energy use in the modification of a shared nest.

Co-operative construction is not limited to adult insects, but is also seen in many insect larvae (Costa 2006). Eastern tent caterpillars, *Malacosma americanum*, collectively build tent structures. Butterflies of this species de-posit 200–400 of eggs near the tip of a cherry twig. In the spring, these eggs hatch nearly simultaneously and the larvae move downward to an intersec-tion on a tree branch. There, together with other caterpillars from nearby branches of the tree, they build a stretchy layer of silk between branches. Construction work is synchronized so that the caterpillars simultaneously produce large quantities of silk to create a new layer (Fitzgerald & Willer 1983). As the layers accumulate the tent begins to dominate the junction between the branches. It is from this central tent that the caterpillars en-gage in a number of other co-operative activities, including pheromone trail foraging (Fitzgerald & Peterson 1983) and group thermoregulation.

The caterpillars' tent often contains genetically unrelated individuals resulting from the merging of groups from batches of eggs laid by differ-ent butterflies. Costa and Ross (2003) looked at how the size and genetic make-up of the colony affected the survivorship and growth rate of in-dividuals within the groups. They found that genetic variation had very little affect on either survivorship or growth rate. However, individuals in larger colonies (100 larvae) grew faster than those in small colonies (30 larvae), although there was no direct effect on survivorship during the study. Again, this would point to a synergism as an explanation of the evolution of co-operation in this case. Once co-operative building is established by evolution, it is sustained by the extra benefits gained from co-operation (see chapter 10).

The relationship between reproductive success and group size brings us back to chapter 2. In another silk-producing arthropod, the social spi-der *Anelosimus eximius*, Aviles & Tufino (1998) showed that individual reproductive success increases with group size (see figure 2.4c). Overall these studies provide support for a mutualistic or synergistic explanation for co-operation in building by groups of insects and other arthropods.

Chambers and Catacombs

The study of pillars and walls has shown that a combined modeling and experimental approach can increase our understanding of how insects produce regular structures. However, obtaining a detailed understanding of the full complexity of social insect nests remains a significant challenge. Figure 7.5 shows nest cavities built by Florida harvester ants. The nest is made up of descending shafts, spiraling downwards with a steepness that increases with depth, from which extend horizontal lobe-shaped chambers. Near the top portion of the structure, these shafts and chambers merge to form highly connected clusters of cavities (Tschinkel 2004). Similar structures are built by fire ants (Halley et al. 2005), *Lasius niger* (Grassé 1959), and other species of ants (Hölldobler & Wilson 1990).

The ant nests in figure 7.5 were dug out by around 5000 ants over 4 to 5 days, during which time they removed about 20 kg of sand and dug to depths of over 3 m (Tschinkel 2004). The question of how such a complex depth-varying structure can be built so rapidly without centralized control remains largely open. Tschinkel suggests that the change of the structure with depth could be due to a combination of a tendency to dig more by workers experiencing lower concentrations of carbon dioxide, which are lower higher up in the nest, and of a tendency of older workers, which prefer lower carbon dioxide concentrations and tend to dig more, to be higher up in the nest. Carbon dioxide could thus provide a template for digging behavior.

The structure of the nests would also suggest some form of positive feedback. In particular, the enhancement of the small notches in the shafts into first circular, then lobe-like chambers is indicative of an increase in digging at points where digging has already occurred. Similarly, regular spacing of the chambers between vertical shafts could result from a form of local activation and long range inhibition similar to that which produces regularly spaced corpse piles in ant cemeteries (box 7.A). One suggestion is that digging ants release a pheromone that increases the other ants' tendency to build (Grassé 1959). Buhl et al. (2005) showed that the amount of sand excavated through time by *Messor sancta* ants is consistent with such a pheromone, although it is less clear whether their model explains the spatial structures.

The experimental approach of Buhl et al. (2005, 2004) provides pointers to how excavated nest structures might be studied further. By placing the ants between glass slides and filming from above they were able to watch the time evolution of the excavations. This allowed them to study the efficiency and robustness of the resulting structures. The ants built a sequence of chambers, connected together by tunnels. Buhl et al. found

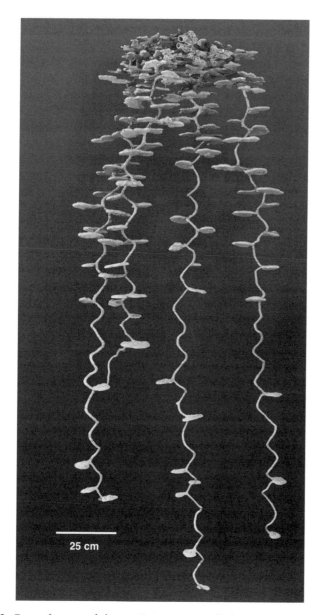

Figure 7.5. Casts of a nest of the ant *Pogonomyrmex badius* (reproduced from W. R. Tschinkel, "The nest architecture of the Florida harvester ant, *Pogonomyrmex badius*," July 2004, *Journal of Insect Science*, 4:21, 1–19, fig. 6c, © Walter R. Tschinkel; photo by Charles F. Badland).

that these networks are robust to disconnections, in the sense that taking away a random tunnel does not lead to disconnection of the network, but still have a low total tunnel length, in the sense that the total amount of tunnel is close to that of a minimal spanning tree. However, while constraining the ants to two dimensions allows the structure they build to be quantified, it is difficult to assess the biological significance of Buhl et al.'s results. In natural conditions the ants build in three dimensions.

Understanding three-dimensional nest structures poses a difficult technical problem. Firstly, in order to characterize a nest the physical structure has to be represented as a graph, with chambers as nodes and tunnels as edges. Perna et al. (2008a) used x-ray tomography of nests of termites of the group *Cubitermes* and then defined measures of elongation in order to classify which parts of the nest were chambers (nodes) and which were tunnels (edges). The second problem in understanding these structures is in measuring their efficiency. Here, Perna et al. (2008b) constructed alternative theoretical graphs, with the same number of edges as the empirically measured graphs but designed to optimize some aspect of the network's topology. The two theoretical graphs they proposed were the random spanning network and maximal betweenness centrality (BC) network. The random spanning network is constructed by first determining the minimum spanning tree for all the nodes of the empirical graph, then adding random edges until the number of edges is equal for the theoretical and empirical graphs. The maximal BC network is constructed starting from a graph consisting of all physically possible edges between the nodes and then determining the shortest path routes between all pairs of these nodes. The edges that occur least often on the shortest path routes are systematically removed, the shortest path routes are recalculated and this process is continued until the same numbers of edges remain in this graph as in the empirically measured network.

Perna et al. (2008b) found that the empirically measured graphs had shorter path lengths than random spanning networks. However, the maximal BC networks had slightly shorter path lengths on average than the empirically measured graphs. These results would suggest that the termites do, at least in part, attempt to minimize the time it takes to move through the network but, as Perna et al. point out, it is difficult to interpret the results without a detailed understanding of the mechanisms involved in nest construction.

Nests of the termite subfamily *Macrotermitinae* are built to optimize the growth of fungi, which these termites cultivate within their nest mounds. The fungi grow best at temperatures of 30 °C and under low concentrations of CO_2. A large termite colony can maintain an internal temperature of close to 30 °C, despite ambient temperature variations of up to 25 °C within a single day and variations of 35 °C throughout the course of the year (Korb & Linsenmair 1998a, 1998b). These constant temperatures are largely a result of the form of the nest structure, with uninhabited nests maintaining a constant temperature of 28 °C (Korb & Linsenmair 2000). The additional temperature in inhabited nests is generated by the fungi and termites.

There is a trade-off between maintaining a temperature of 30 °C and low concentrations of carbon dioxide (Korb 2003). In a series of observations and experiments, Korb and Linsenmair have compared the mound structures built by *Macrotermitinae bellicosus* living in different sized colonies in different environments. In the warm Savannah, large colonies maintain 30 °C nest temperature and construct a cathedral-shaped mound with ridge-like chimneys. These chimneys efficiently transport carbon dioxide out of the nest. In cooler, forested areas the termites build more dome-like mounds with reduced surface complexity and less effective ventilation. These maintain higher temperatures, although lower than the optimal 30°C, but are less effective at circulating air. When the shade is removed from these mounds the termites surface complexity increases (Korb & Linsenmair 1998a).

It is likely that termites regulate their building behavior in response to temperature and carbon dioxide concentration. Leaf-cutter ants, which also cultivate a fungi garden that requires constant ambient temperature and high humidity, show an increased tendency to build in areas they detect the flow of dry air (Bollazzi & Roces 2007). In general, social insects can use naturally occurring features of their environment combined with simple behavioral rules to construct nests with highly efficient thermoregulatory properties (Jones et al. 2007; Theraulaz et al. 2003).

Foraging Networks

In chapter 3, I discuss one of the best known examples of indirect communication between social insects, namely ant pheromone trails. That chapter was mainly concerned with the organization of foraging at only one or two food sources. Figures 7.6a, 7.6d and 7.7 give examples of the foraging networks of various species of ants under natural conditions. These networks are typically dendritic in form (Hölldobler & Möglich 1980; Hölldobler & Wilson 1990). Each trail starts from the nest as a single thick pathway out of the nest. This "trunk" splits first into thinner branches, and then peters out as the distance from the nest increases, into twigs, often barely distinguishable in the undergrowth. While sharing this dendritic form, there are between species differences in the time for

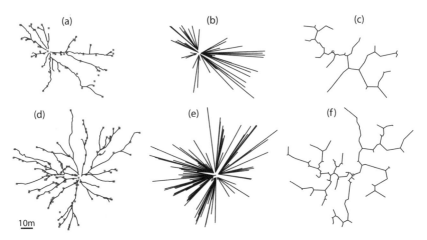

Figure 7.6. The trail networks constructed between the nest and trees by (a, d) two wood ant colonies (*Formica rufa aquilonia*); compared to (b, e) the star graph connecting each tree to the nest; and the approximate Steiner tree that minimizes total exposed trail. (Figure drawn by Jerome Buhl.)

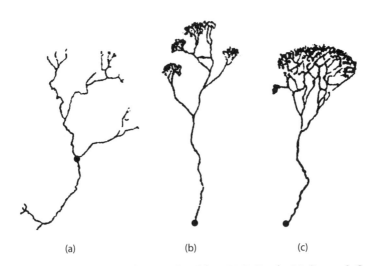

Figure 7.7. Ant foraging networks (reproduced from N. R. Franks, N. Gomez, S. Goss, & J. L. Deneubourg, "The blind leading the blind in army ant raid patterns: Testing a model of self-organization (Hymenoptera: Formicidae)," September 1991, *Journal of Insect Behaviour* 4: 5, 583–607, fig. 1, © Springer-Verlag). Short lived raid networks constructed by three different species of army ant, *Eciton harnatum*, *E. rapax*, and *E. Burchelli* (from left to right), each covering some 50 × 20 m.

which trails persist and the mechanisms used in their construction. Here I separate the types of trails into two categories, long-lasting transport networks or short-lasting raid networks.

Transport Networks

Wood ants (Chauvin 1962; Rosengren & Sundström 1987), leaf-cutter ants (Shepherd 1982; Vasconcellos 1990; Weber 1972), and harvester ants (Azcarate & Peco 2003; Detrain et al. 2000; Hölldobler 1976; Hölldobler & Möglich 1980; Lopez et al. 1994) produce physical trails that can last from several weeks to months, and in some cases endure the winter hibernation period (Fewell 1988; Hölldobler & Möglich 1980; Rosengren & Sundström 1987; Weber 1972). Workers clear trails of vegetation and debris and sometimes construct walls or tunnels around them (Anderson & McShea 2001b; Kenne & Dejean 1999; Shepherd 1982) to form highways along which large numbers of ants are able to travel quickly to food. Both pheromone trails and the clearing of vegetation provide mechanisms for the formation of trail networks.

Buhl et al. (2009) investigated the shape of 11 nests of the wood ant *Formica aquilonia*. The trees at which the ants forage constitute the *vertices* of the network, while the trails are the network's *edges*. The central vertex of the network is the ant's nest. The networks were characterized in terms of two components of efficiency, the route factor and total edge length (Gastner & Newman 2006). The route factor is the average of the distance between each vertex and the central vertex when traveling within the network divided by the direct Euclidean distance (i.e., the distance as the crow flies) between the two vertices. A low route factor corresponds to a short travel time between nest and food source. The total edge length is the sum of all the edges in the measured network. Since trails must be kept clear in order that ants flow quickly along them, networks with low total edge length are more efficient in terms of the "maintenance" required.

Figure 7.6 compares the networks constructed by two wood ant colonies with those that would minimize route factor and those that would minimize total edge length. The network that minimizes route factor has a star-like shape, with a direct edge between every vertex and the nest (figure 7.6b, e). As a result this network has a very high total edge length. Conversely, the network that minimizes total edge consists of clusters of triangles and long sprawling paths connecting nearby vertices, thus having a large route factor (figure 7.6c, f). The ants obtain a compromise between the two modes of efficiency. The average route factor was 1.13 (compared to a theoretical minimum of 1) with the largest colony having a route factor of 1.36. The total edge length for the 11 measured colonies was between 1.01 to 1.73 times that of the corresponding networks that minimize the total edge length. Thus the ants produce networks that provide fast travel times from the center of the nest without requiring direct paths to every food source.

How are these networks built? Acosta et al. (1993) provide an intuitive argument for how branches arise in long term trails. They argue that when a forager finds a resource at a point perpendicular to an established trail, it returns to the established trail leaving pheromone as it goes. This forager, and subsequent foragers that have followed the pheromone to the resource, will have a tendency to walk towards the nest thus diverting the newly formed trail such that branching angle decreases. Eventually a Y-shaped branch emerges with the branching point some distance away from the two food sources.

A variation of Acosta et al.'s argument is formalized in a mathematical model in box 7.B. In this model, two resources are simultaneously made available to the ants at equal distances from the nest. Two assumptions are made about the movement of the individuals: that they have a long range navigation, which allows them to move roughly in the direction of the food source; and that they are locally attracted to paths taken by others. The relative importance of these two rules is controlled by model parameter α. When α is small the individuals ignore the paths taken by others and when α is large they prefer to follow paths even if they do not lead directly to the target. Figure 7.8 gives examples of the outcome of simulations of this model for various values of α. The resulting path system changes from a V- to a Y- and then to a T-shape as the preference for path following increases.

Since wood ants are able to navigate over long distances between food and their nest without the use of pheromones, the above model is a plausible description of how their network arises. Further work is needed to establish whether such a model can account for the patterns observed in their real networks.

Raid Networks

Army ant species (Franks 1989; Schneirla 1971; Topoff 1984), as well as *Leptogenys processionalis* (Ganeshaiah & Veena 1991), and *Pheidologeton diversus* (Moffett 1988) all form swarm raid trails that last for a couple of days or less. These trails result from strong positive feedback from recruitment pheromones. The short-lasting raid patterns by the army ant (Deneubourg & Goss 1989; Franks 1991) and the predatory ant *L. processionalis* (Ganeshaiah & Veena 1991) have been measured in detail. These networks at first expand as the ants search for food and then tend

Box 7.B Active Walker Model

The active walker model of Helbing et al. (1997a, 1997b) makes two assumptions about the movement of individuals through a shared environment. Firstly individuals move towards a target and secondly they are locally attracted to the paths taken recently by others. Here I develop a simplified version of the Helbing et al. model, based on these assumptions.

Consider an individual traveling to a particular destination at position (x_s, y_s). We assume the individual is confined to a square lattice so that, given its current position (x_t, y_t), it must decide on each time step whether to go left (i.e., $x_{t+1} = x_t - 1$ and $y_{t+1} = y_t$), right (i.e., $x_{t+1} = x_t + 1$ and $y_{t+1} = y_t$), up (i.e., $y_{t+1} = y_t + 1$ and $x_{t+1} = x_t$) or down (i.e., $y_{t+1} = y_t + 1$ and $x_{t+1} = x_t$). If the destination is to the left of the individual (i.e., $x_t > x_s$) then the probability of going left is set to be proportional to

$$l = \frac{x_t - x_s}{|x_t - x_s| + |y_t - y_s|}(1 + \alpha G_t(x_t - 1, y_t))^2 + k,$$

otherwise (i.e., $x_t < x_s$) we set $l = k$. Similarly, if the destination is below the individual (i.e., $y_t > y_s$) then the probability of going down is set to be proportional to

$$d = \frac{y_t - y_s}{|x_t - x_s| + |y_t - y_s|}(1 + \alpha G_t(x_t, y_t - 1))^2 + k,$$

otherwise (i.e., $x_t < x_s$) we set $d = k$. Similar calculations can be made for going up, u, and right, r. The probability of going, for example, left is then normalized, to be equal to $P(x_{t+1} = x_t - 1, y_{t+1} = y_t | x_t, y_t) = l/(l + r + u + d)$.

The variable $G(x, y)$ measures attractiveness of the point (x, y). We can think of $G(x, y)$ as being the amount of pheromone at that point or the extent to which the ground is trodden down by previous walkers. In the absence of pheromone, i.e., $G(x, y) = 0$, or if pheromone is ignored, i.e., $\alpha = 0$, the individuals perform a random walk biased towards the destination. Smaller values of k will result in less randomness in the walk. The paths taken by 50 individuals for this case are shown in figure 7.8a.

(Box 7.B continued on next page)

If we assume that each individual increases $G(x, y)$ (i.e., tramps down the ground or leaves pheromone) at its current position but that $G(x, y)$ degrades (e.g., the grass grows back or the pheromone evaporates), then we have the following equation for G

$$G_{t+1}(x,y) = \lambda G_t(x,y) + \kappa(1-\lambda)I_t(x,y),$$

where $I_t(x,y) = 1$ if an individual is at point (x, y) at time t, κ is the amount of "pheromone" deposited by one individual and λ is the evaporation rate. Figure 7.8b–d shows the paths taken by 50 individuals simultaneously walking on a shared path system according to the above rules for different values of α. As α increases it becomes favorable for individuals to use shared paths and the movement of individuals becomes more focused along these paths.

to contract once resource items are located and foraging is focused on these resources. As such, these exploratory networks provide a trade-off between a minimization of the cost of travel and a maximization of the area over which the ants search for prey items.

As with nest construction, Deneubourg and Franks provided the first combined modeling and experimental approach to the study of raiding networks. Using computer simulations, Deneubourg et al. (1989) showed that the networks created by army ants during a raid could be reproduced by the simple rules for pheromone laying and following established in double bridge experiments (chapter 3). Their model made three key assumptions: (1) ants lay pheromone both on the way out to and the way back from a food source; (2) at a branch in the pheromone trail the probability the ants take the left branch is

$$\frac{(x+k)^{\alpha}}{(x+k)^{\alpha} + (y+k)^{\alpha}},$$

where x is the amount of pheromone on the left branch and y the amount on the right branch, and k and α are constants (see box 7.B for a similar model; also chapter 3, equation 3.1); and (3) the ants' speed is proportional to pheromone concentration. Franks et al. (1991) confirmed experimentally that assumptions 2 and 3 held for the army ant *Eciton burchelli* (assumption 1 was already known to hold).

Simulations of the model showed that these three aspects of the ants' behavior were sufficient to generate patterns similar to the raid patterns seen in real ant colonies (figure 7.7). This was quite a remarkable result

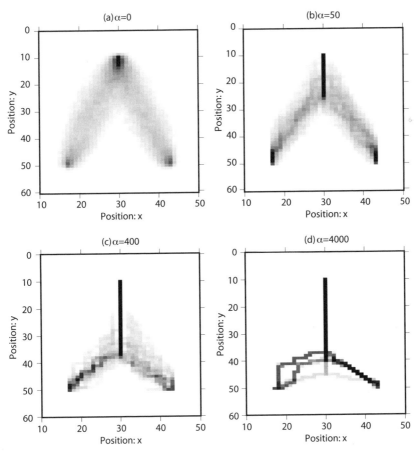

Figure 7.8. Simulation of the simplified active walker model (box 7.B). In the simulation 50 individuals start at the nest, i.e., $(x_0, y_0) = (30, 10)$; 25 of the individuals have food target $(x_s, y_s) = (15, 50)$ and the other 25 have food target $(x_s, y_s) = (45, 50)$. The walkers then move and the trail is updated according to the rules given in box 7.B. Once walkers arrive at the food, i.e., $(x_t, y_t) = (x_s, y_s)$, their target is updated to be the nest, $(x_s, y_s) = (30, 10)$. Correspondingly, when walkers arrive at the nest their target is updated to one of the two food sources (chosen at random). The simulation was run for 150,000 time steps and the positions of individuals during the last 50,000 time steps recorded. The panels show the average number of visits by walkers to each of the positions over these last 50,000 time steps. The simulation parameter values were $k = 0.25$, $\lambda = 0.99$, $\kappa = 0.8$ and (a) $\alpha = 0$, (b) $\alpha = 50$, (c) $\alpha = 400$, and (d) $\alpha = 4000$.

because, while it was known that models could predict foraging patterns in double bridge experiments, these simulations showed that the laboratory results could be scaled up to large-scale foraging patterns in the field. Another prediction of the model was that the differences in the raid patterns of different species (seen in figure 7.7) may not result from differences in their behavior, but rather in differences in the distribution of their food. By manipulating the food distribution, Franks et al. (1991) showed that changes in food distribution did lead to large-scale changes in foraging patterns. While these changes were different than those predicted by the model, this study challenged the perspective that differences in the patterns generated by insect societies are a consequence of the evolution of different behavior in different environments. Rather, the same simple rules can self-organize into different collective patterns in different environments without the "fine tuning" of evolutionary forces.

Man-made Networks

We don't have to look much further than under our own feet to find transport networks that are self-organized. The model in box 7.B was originally proposed in the context of human trail systems in green areas, such as business parks and between university buildings. Often in these areas we can observe "shortcuts" taken across the lawns separating buildings, resulting in trails of downtrodden grass. The two assumptions underlying the model in box 7.B, that there is a long range attraction to a target and a local attraction to previously used paths, are consistent with how people move when crossing grass. Helbing et al. (1997a, 1997b) showed that a model based on these assumptions was sufficient to generate the observed network of shortcuts. The model reproduced a number of more subtle features of these networks, such as "detour systems," which are a compromise between direct routes to all targets and a system that shares common routes. As with ant trails, the resulting detour system is a compromise between one that minimizes route factor and one that minimizes total edge length.

We can think about the detour system in terms of the motives of the individuals involved in its construction. From the point of view of each "selfish" individual, their payoff is maximized if a direct route emerges between their starting point and their destination. However, as a result of the simultaneous interaction of large numbers of individuals with different starting point/destination combinations a compromise emerges. This compromise has lower route factor than the Steiner tree and only slightly higher route factor than a square drawn between all four points. This is interesting because the route factor is a measure of group "fitness," rather than individual "fitness," i.e., it is determined by the average travel time across all participating individuals. As such, these detour systems may provide a guideline for urban planners wishing to avoid "selfish" people taking shortcuts across lawns.

Many human transport systems and supply networks are designed by a central planner. These planners have an interest in minimizing the average travel time of individuals across the whole system, while ensuring that the systems are not too expensive in terms of maintenance. Gastner and Newman (2006) looked at the route factor and total edge length of a sewer system, a rail network, and two gas supply lines. Like the wood ant networks, human networks obtained a compromise between the two efficiency measures. Comparing the results of Gastner & Newman with those of Buhl et al. shows that the transport networks of ants and humans have similar average route factor and total edge length.

A key difference between construction by ants and by humans is the possibility of central planning by the latter species. Gastner and Newman (2006) propose a simple algorithm for the growth of transport networks, based on a central planner, which reproduces many of the features of real networks. However, central planners are by no means a necessity for construction by humans. Many of the world's largest cities arose through local, rather than global planning. Makse et al. (1995) proposed that cities were constructed according to the following two principles: (1) the probability of development decreases exponentially with distance from the center of the city; and (2) future development is correlated with past development, such that the probability of a particular site being developed increases with the existence of nearby, already developed sites. The resulting model is a spatial variation of preferential attachment models discussed in chapter 2, box 2.C.

Makse et al. (1995) compared their model to the structure and development of Berlin and London. In particular, they looked at the area distribution of the towns surrounding the city center. For both cities, this distribution followed a power law with an exponent slightly larger than 2. This exponent was consistent with a particular parameterization of their model, in which new developments were strongly correlated with previous ones. These results should be interpreted with the care required when dealing with mechanistic interpretations of power laws. A power law scaling does not in itself imply a particular mechanism (see chapter 2 for more discussion of this point). The Makse et al. (1995) study does, together with several other similar studies, suggest new ways of understanding, analyzing, and predicting the growth of cities using models of self-organization (Batty 2008).

A particularly striking study of this type is Kuhnert et al.'s (2006) study of urban supply networks. They looked at how the number of restaurants, doctors, pharmacies, post offices, gas stations, car dealers, and other services per person change with the size of the town or city they were found in. Those services provided by the government or deemed essential to modern living, such as power station output, pharmacies, and post offices, increased in proportion to the size of towns. The increase in the number of gas stations and car dealers with population size was sub-linear, suggesting an economy of scale for large cities. Conversely, the proportion of restaurants, museums, and theaters per person increased super-linearly. Larger towns saw higher concentrations of entities that supply entertainment and non-essential social needs. This may be because of the increase in disposable cash and higher education of individuals living in large cities. Bettencourt et al. (2007) found that wages, gross domestic product (GDP), and employment in "creative" jobs also increased with city size, as did the overall pace of life.

Manmade networks have been the focus of intense study in recent years. Indeed, the study of networks is not limited to physical entities such as transport, supply, or computer networks. One of the most studied types of network is social networks, where the connections are created by contacts between individuals. There are several excellent reviews and books on social networks, and it is this vast literature on the subject that constrains me from writing more about it here (Barabasi 2003; Newman 2003; Strogatz 2001; Watts & Strogatz 1998). One of the major challenges that lie ahead of us is linking together characterization of the networks within which individuals interact (as is often studied in social network theory), with dynamical models about how individuals make decisions (which is more often the focus of work on collective behavior).

— Chapter 8 —

Regulation

At any time during the working day, I can get up from my desk, walk down to the cafeteria, and find a container full of hot coffee from which I can pour myself a cup. The fact that the coffee is there waiting for me is not a consequence of careful preparation for my arrival by the cafeteria staff. I could go across to the next building, where I have never been before, walk into the basement café and sitting there waiting for me would be a similar container also filled with coffee. Not only coffee, but food, clothes, houses, and everything else I need for modern living appears perfectly regulated for my needs. When I want something it is there waiting for me. When things are not readily available—for example, nice houses being difficult to find; the supply of the latest game machine running out just before Christmas; or no coffee available in the cafeteria at 11am— this becomes the subject of intense discussion about how the suppliers should act to rectify the situation. These situations are then often quickly rectified, as demand increases and supply decreases, or new businesses appear to fill the gaps in the market. Consumers expect and receive supply that is regulated to suit their needs.

Regulation of supply and demand does not require central planning by me or anyone else. I do not have to call down to the café in advance and ask them to switch on the percolator; the cafeteria owner does not have to know when the next boat of coffee beans is coming from South America; and the shipping agent does not need to check that new plants are already in the ground for the next year's crop. Through a series of local economic interactions I am provided with a regular supply of coffee. It was both an amazement and an understanding of how unmanaged and unguided activities of humans produce equilibriums in supply and demand that led Adam Smith to describe the economy as if being guided by an invisible hand. For me, this amazement and understanding is best expressed in the opening sections of Thomas Schelling's book *Micromotives and Macrobehavior* (1978). Schelling identifies the importance of the fact that "the market works" not just in economics, but in all forms of collective behavior.

This chapter is about when markets work and when they do not work. More broadly it is about systems that involve regulatory feedback. Regulatory feedback, which is also referred to as negative feedback, is when a system responds in an opposite direction to a perturbation. For example, if a café has many visitors, the café employees make more coffee. If queues become too long customers stay away. The individual agents, the employees and the customers, behave in a way that eventually leads to an equilibrium queuing time. Regulatory feedback usually, though not always, performs a balancing act that stabilizes systems, bringing them to equilibrium. When it works effectively, regulatory feedback balances supply and demand, not only in our own society but also in the workings of other animal societies.

Co-operative Regulation

Workers in insect societies often share a common goal of achieving a stable response to their environment, usually in the form of some optimal balance in their intake of water, sugar, and other resources. For example, honeybees need to construct comb in which to store incoming nectar. Deciding when to build this comb poses a challenge for the bees, because construction is energetically costly and nectar intake is highly variable. To address this challenge, the bees only begin construction of new comb when there is both a high rate of nectar flow into the colony and the available comb drops below a threshold level (Kelley 1991; Pratt 1999). This strategy ensures that comb is available over a wide range of foraging conditions, even when sudden food 'bonanzas' become available (Pratt 1999, 2004).

Simple rules, similar to those of building when nectar flow is high and comb availability is low, are used to regulate a whole range of tasks within honeybee colonies (Seeley 1995). A major challenge in understanding this regulation is pinpointing how individual bees gather information about which tasks need to be performed. How do bees, with only a limited experience of the overall status of the colony, know when to begin a particular task? In comb building these rules remain unknown, although experiments on the construction of drone cells (comb for rearing male bees) suggest that individual builder bees may monitor the proportion of empty cells (Pratt 1998a, 1998b). In general, a number of theoretical models have emphasized how efficient division of labor, i.e., allocation of worker bees among tasks, can emerge from individuals using local information about their environment (Anderson & Ratnieks 1999, 2000; Bonabeau et al. 1998a; Franks & Sendova-Franks 1997; Gordon 1996; Gordon et al. 1992). Individuals change their propensity to perform a particular task in response to the information they collect about the state of the environment and as a result the colony as a whole regulates its allocation of workers to tasks.

The entrance to a honeybee colony, often referred to as the dancefloor, is a market place for information about the state of the colony and the environment outside the hive. Studying interactions on the dancefloor provides us with a number of illustrative examples of how individuals changing their own behavior in response to local information allow the colony to regulate its workforce (Seeley 1995, 1997). For example, upon returning to their hive honeybees that have collected water search out a receiver bee to unload their water to within the hive. If this search time is short then the returning bee is more likely to perform a waggle dance to recruit others to the water source (Lindauer 1954). Conversely, if this search time is long then the bee is more likely to give up collecting water (Kuhnholz & Seeley 1997). Since receiver bees will only accept water if they require it, either for themselves or to pass on to other bees and brood, this unloading time is correlated with the colony's overall need of water. Thus the individual water forager's response to unloading time (up or down) regulates water collection in response to the colony's need.

Similar regulatory interactions also determine how honeybees increase the overall level of activity within the colony (Seeley et al. 1998); decide when to build drone comb (Pratt 1998b); decide whether to scout for new or exploit known food sources (Beekman et al. 2007); decide when to collect pollen (Fewell & Winston 1992); and decide whether to nurse younger or older developing larvae (Schmickl & Crailsheim 2002). Seeley (1995) gives an authoritative review of these and many other regulatory feedback mechanisms within the honeybee colony. One of the most striking examples of regulatory feedback is seen in nectar processing. When returning from a successful foraging trip, a forager bee performs either a waggle dance or a tremble dance. The choice of dance depends on the time it has searched for a bee to which to unload the nectar it has found. Waggle dances result in the recruitment of more foragers, reflecting the colony's need for more nectar, while tremble dances result in the recruitment of more nectar receivers reflecting an increased influx of nectar. Combined, these two regulatory feedbacks ensure that nectar flow is not delayed by a shortage of either foragers or receivers.

In many ant species, contacts between workers are used to regulate the division of labor. Gordon et al. (1993) showed that ants regulate their degree of antennal contacts, aggregating more when density was low and less when density was high. While these experiments were carried out in rather artificial conditions, Gordon hypothesized that contact rates could provide a general explanation of how workers regulate their division of labor (Gordon et al. 1992). Experimental evidence now supports this hypothesis. For example, the safe return of morning patrollers to a red harvester ant nest results in an increase in the number of foragers leaving the nest to collect food (Gordon 2002; Greene & Gordon 2007a, 2007b). Similarly, ants that encounter nestmates engaged in refuse pile maintenance are more likely to engage themselves in refuse work (Gordon & Mehdiabadi 1999). In leaf cutter ants there is an additional regulation of refuse work whereby non-refuse workers aggressively force contaminated workers to remain and work in the garbage area (Hart & Ratnieks 2001, 2002).

Regulatory feedback is not necessarily a result of direct contacts between individuals, but can also be mediated through interactions with resources and/or pheromones. Maileux et al. (2000) found that if *Lasius niger* ants find food that allows them to ingest a desired volume they leave pheromone trail and recruit nest-mates, but if they cannot obtain this volume they return to the nest without recruiting. This simple rule of thumb prevents recruitment of an excess of foragers to a site with only small amounts of food or where food has been depleted. This rule is extended such that when food is made up of small sub-units the ants scout locally around their first local discovery and then recruit to this site if there exist other nearby sub-units (Mailleux et al. 2003b). Other species of ants may lay pheromones to prevent other ants going to areas they have already explored and failed to find food. For example, Robinson et al. (2005) found evidence that Pharaoh's ants mark unrewarding foraging paths.

In all these examples of regulatory feedback, individuals do not have global knowledge of the state of the colony or the distribution of food. Rather they have made their own samples of the available food or the time they had to wait to unload and regulated their own behavior in an appropriate way. For example, it may well be that a particular forager has by chance a short unloading time for a commodity that is not needed within the colony. However, if a large number of bees are simultaneously attempting to unload this same commodity, then on average their unloading time will be long and the number of individuals collecting this commodity will be down-regulated. Local sampling, performed in parallel by large numbers of individuals, allows the colony to accurately tune its average response to changes in the environment.

Over-compensation and Chaos

In the above examples we can distinguish two types of regulatory feedback: passive and active. Passive regulatory feedback involves individuals adjusting their internal probability of performing some action based on their own success. For example, the decision to abandon the utilization of a food source is a function of an individual's own success in obtaining food there. Active regulatory feedback involves individuals producing a cue or a signal that in turn changes the probability of other individuals performing some action. For example, negative and positive pheromones are signals that respectively down- or up-regulate the number of ants taking a particular path to food. Active regulatory feedback may be either cue- or signal-based. For example, the decision by a bird to search for food in a particular place may be copied from that of other individuals (i.e., cue-based), even though the individual making the decision was completely unaware of being copied. I classify this as active feedback, even though there was no deliberate communicative action on the part of the copied individual.

Passive regulatory feedback usually results in a stable equilibrium, while active regulatory feedback can result in over-compensation. Box 8.A describes a model that illustrates this point, showing how active regulatory feedback can produce oscillations whereby the population never reaches equilibrium and can even become chaotic. In the model in box 8.A there are three factors that are required to produce over-compensation and chaos in regulatory feedback. The first is that feedback is active, if individuals simply retire from a food source when it becomes overcrowded and return independently from each other to assess whether it is exploitable then collectively they will not over- or under-shoot the equilibrium. The second factor is that information is local. Individuals that have sampled a single food unit cannot determine whether their experience reflects the overall state of the resource. If individuals were able to assess the entire resource then they would have a fuller picture of the effects of recruitment. The last factor is that there is a time delay between the observation and the regulatory response. The generation of instability and chaos depends on discrete time steps. If these are taken away, the oscillations are dampened out.

Insect societies do sometimes over-compensate for changes in the environment. For example, when starved ants are offered food they typically recruit strongly to it at first, leading to over-crowding at the food source (Mailleux et al. 2003b; Pasteels et al. 1987). Once the food source is overcrowded recruitment is reduced, but often not until after some individuals have arrived to find the food overcrowded (Detrain & Deneubourg 2006). Given that such over-compensation can lead to chaotic oscillations, why is active regulatory feedback so common in the interactions of social insects? A first answer to this question lies in the advantages of active feedback when a system is a long way from equilibrium. In chapter 3 we saw that positive feedback is highly effective in transmitting new information. In dynamically changing environments, positive feedback can communicate changes in the environment without requiring every individual to experience the change itself.

Box 8.A Over-compensation and Chaos

Consider individual agents, each of which aims to exploit a resource, such as a flower patch or a Thursday night music bar.
Assume that the resource is composed of n smaller units, e.g., individual flowers within a flower patch for foraging honeybees or
chairs in a bar, and further assume that each of these units may
only be exploited by one individual at a time. The division of the
resource into smaller units means the information individuals obtain about the resource is local. Individuals sample one unit and
have to decide what to do next on the basis of their experience at
this unit. We assume that within the resource, individuals choose
units entirely at random. This assumption implies that, provided n
is reasonably large, the number of workers choosing a particular
food unit is Poisson distributed, i.e., the probability that k individuals choose a resource unit is

$$p_k = \frac{(x_t/n)^k}{k!} e^{-x_t/n}.$$

See, for example, Brännström & Sumpter (2005) for a derivation
of the Poisson distribution in this way.

I now discuss separately two ways in which individuals might respond to their experience in deciding whether or not to go to visit
the resource. Under passive regulatory feedback, we assume a constant flow α of individuals who spontaneously decide to test the
resource. We assume that if two or more individuals choose the
same unit then they both conclude the resource is overcrowded
and decide not to try to exploit it on the next time step. We can
then write the equations for the number of individuals visiting the
resource x_t through time t as

$$x_{t+1} = \alpha + p_1 n = \alpha + x_t e^{-x_t/n} \equiv f(x_t) \qquad (8.A.1)$$

Figure 8.1 plots a time series of iterations of equation (8.A.1) along
with a cobweb diagram showing how these iterations converge on
a stable equilibrium, x_*, which satisfies $x_* = f(x_*)$. While we cannot
write down a simple closed form expression for x_*, it is relatively
straightforward (although I do not do it here) to show it exists and
lies between 0 and n. Differentiating f with respect to x we get

$$f'(x) = (1 - x/n)e^{-x_t/n}.$$

Thus, $-1 \le f'(x) \le 1$ for all $0 \le x \le n$ and thus the slope of $f(x)$ near to the equilibrium is less than one. This observation implies that the equilibrium is stable (for more details of the methodology underlying these conclusions see, for example, Strogatz 1994). In words, passive regulatory feedback results in the number of individuals at the resource stabilizing at the unique value of x_*, for all possible values of α and n.

Active regulatory feedback includes some form of recruitment to the resource, e.g., pheromone trails left by ants, dances by bees, or spreading the word to friends by humans. Assume that individuals use the following two simple rules: (1) if an individual chooses a resource unit that no other individual chooses they conclude that the resource has excess capacity and recruit to the resource an average of $b - 1$ other individuals who weren't previously at it; alternatively, (2) individuals choosing a unit chosen by another individual conclude that the resource is over exploited and decide not to come back on the next time step. For large n these rules give the following equation for the number of individuals visiting the resource x_t through time t as

$$x_{t+1} = bp_1 n = bx_t e^{-x_t/n} \equiv g(x_t). \qquad (8.A.2)$$

Figure 8.2 plots a time series of iterations of equation (8.A.2) along with a cobweb diagram of how consecutive populations change for different values of b.

For small b, iterations of equation (8.A.2) converge on a stable equilibrium, $x_* = n\ln(b)$, which satisfies $x_* = g(x_*)$. However, differentiating g and evaluating at x_* gives

$$g'(x_*) = 1 - \ln(b).$$

If $\ln(b) > 2$ then the slope at the equilibrium is less than -1, implying that the equilibrium is not stable when $b > e^2$. Figures 8.2(a) and (b) give an example with $b = 6$. For b slightly larger than e^2 the population will oscillate around the equilibrium $x_* = n\ln(b)$ but never come to rest there. An example of this is seen in figure 8.2(c) and (d) for $b = 8$. When the population at the resource

(Box 8.A continued on next page)

179

is slightly below the equilibrium, recruitment overcompensates and the population becomes larger than the equilibrium. When the population is larger than the equilibrium abandonment again overcompensates and moves to below the equilibrium. The cycle of overcompensation continues indefinitely.

Things become even less stable as b increases further. Figures 8.2(e) and (f) give an example with $b = 20$. Here the population oscillates wildly, with populations sometimes large and at other times small. Far from being regulated, the population never settles to anything near to an equilibrium. In fact, equation (8.A.2), more widely known as the Ricker map, is an example of a chaotic dynamical system (May 1976; Strogatz 1994; Sumpter & Broomhead 2001). If we were to start with two similar but slightly different population sizes, then within a few generations these differences would become amplified in a way that would make it impossible to reliably predict future population sizes. Active regulation with strong feedback leads to chaotic population dynamics. See Sumpter & Broomhead (2001) for further investigation of this model.

Even when positive feedback is used to actively up-regulate the number of individuals engaging in a particular task, under most natural conditions equilibrium is reached. Indeed, the three factors that are needed for over-compensation to occur are unlikely to be present simultaneously. In particular, positive feedback is not usually particularly strong within many insect societies. Seeley (1995) emphasizes the economy of communication within honeybee colonies. His picture of a honeybee colony is one of an ensemble of individuals that rather infrequently exchange information with each other. Each individual adjusts its behavior in response to changes in their shared environment. Where positive feedback is strong, for example in ant foraging, other passive regulatory mechanisms, such as retirement in response to overcrowding, operate to damp down initial over-compensation.

Selfish Regulation

Honeybees and other social insects regulate variables in which they share a common interest, usually those that aid the successful reproduction of the queen. However, regulation is not limited to situations where

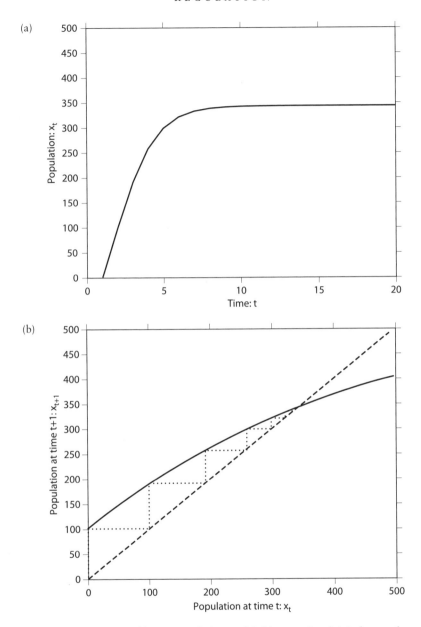

Figure 8.1. Illustration of how a population modeled by equation 8.A.1 changes through time. The parameters are $\alpha = 100$ and $n = 1000$. The time series (a) shows that the population of individuals exploiting the resource equilibrates. The cobweb diagram (b) provides a plot of equation 8.A.1 (solid line), as well as the line $x_t = x_{t+1}$ (dashed line). The point x_*, which satisfies $x_* = f(x_*)$, is the equilibrium population of individuals exploiting the resource. The dotted line shows how consecutive iterations of f move towards this equilibrium.

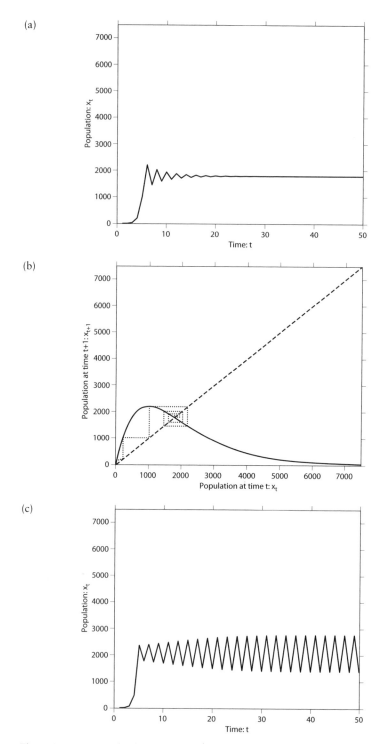

Figure 8.2. Numerical solution of how a population modeled by equation 8.A.2 changes through time. The parameters are $n = 1000$ in all plots with (a, b) $b = 6$, (c, d) $b = 8$, and (e, f) $b = 20$. The time series (a, c, e) shows how the population of individuals exploiting the resource changes through time. The cobweb diagrams (b, d, f) provide a plot of equation

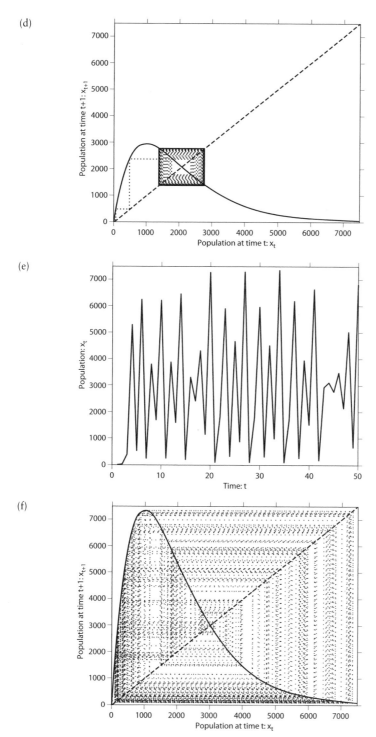

8.A.1 (solid line), as well as the line $x_t = x_{t+1}$ (dashed line). The point x_*, which satisfies $x_* = f(x_*)$, is the equilibrium population of individuals exploiting the resource. The dotted line shows consecutive iterations of f. When $b = 6$ the population equilibrates, when $b = 8$ the population cycles periodically, and when $b = 20$ it is chaotic.

individuals have a common interest in a particular variable reaching equilibrium. In a queue for coffee a particular equilibrium queue length is not necessarily regulated by the café's management or the customers, it emerges from their respective and different aims.

Microeconomics is primarily concerned with finding the equilibrium price of commodities, where supply equals demand. Given supply and demand curves as a function of price, the equilibrium market price is the point at which these two curves cross. Schelling (1978) emphasizes that the fact that an equilibrium is reached should not be confused with the idea that the equilibrium is "good" or in some way optimal for those involved. At a café, it may be that regulatory feedback has led to people staying away from the café because the queues were too long. These people staying away make queues shorter for everyone else, but it is perhaps not the optimal solution for the average customer or the café management. Much of the theory of economics is about designing markets that reach an equilibrium that optimizes some criteria, be it maximizing economic growth or minimizing carbon emissions. Microeconomics is thus a powerful tool for solving these problems and forms the backbone of economic theory and practice (Fama 1970, 1991) and is the subject of a large number of textbooks (Krugman & Wells 2004; Perloff 2007).

The theory of microeconomics is grounded in the idea of the rational agent, attempting to maximize its own utility, which usually but not always means financial benefit. The interactions between rational agents provide a regulatory feedback that brings the market to equilibrium. If an item is over-priced, demand falls and as a result so does price. If the item is under-priced then demand exceeds supply and the price increases. Regulatory feedback on the price of a commodity brings supply and demand to equilibrium. This mechanism is not only theoretically grounded, but is consistent with our everyday experience. That prices in supermarkets are stable from week to week is testimony to market equilibrium. A store raising or lowering its prices by large amounts relative to those of its competitors would not remain in business for very long.

While prices of many everyday commodities may be stable, many economic systems are a long way from equilibrium. The growth of western economies over the last 100 hundred years has been characterized by large fluctuations over the time scale of years, where periods of accelerated growth are proceeded by recession and then a return to growth (Ball 2004; Ormerod 1998). Furthermore, variations in the price of financial markets on shorter time scales of one minute to one week are power law distributed (Mantegna & Stanley 1995; Potters et al. 1998; Stanley et al. 2001). Rather than prices being normally distributed noise around an equilibrium, as would be expected in a stable market, large fluctuations in prices occur on a daily basis. Financial markets can only be considered in equilibrium if viewed on the time scale of months. Both longer and shorter time scales show non-normally distributed fluctuations.

The long time scale fluctuations, often referred to as the business cycle, pose a challenge to the rational agent hypothesis for regulation. If there exist predictable cycles in economic growth then rational agents would be able to exploit these cycles for their own financial gain, thus reducing and eventually removing the cycles. There are two main schools of thought aimed at resolving this paradox (Vercelli 1991). The first of these, following the microeconomic theory of rational agents, sees the business cycle as something that is externally generated by, for example, lags in the time between clusters of technological innovations and the capital generated by these innovations (Kydland & Prescott 1982; Lucas 1975). The alternative school of thought sees the business cycle as an intrinsic property of economic activity, explaining it either in terms of macroeconomic variables (Krugman 2005; Maynard Keynes 1936) or in terms of agents with limited information or, what is known by economists as bounded rationality (Arthur 1994; Conlisk 1996; Ormerod 1998).

A full discussion of how and why the business cycle arises is well beyond the scope of this book. For my purposes, the important observation here is the link between models of cycles in economics and mechanistic explanations of regulatory feedback. Arthur (1994, 1999) proposed a toy economic model, which he called the El-Farol bar problem, for investigating how different mechanisms might or might not lead to equilibrium. Consider a bar that has a music night every Thursday. We can define a payoff function, $f(x)$, which measures the "satisfaction" of individuals at the bar attended by a total of x patrons. One example of such a payoff function is $f(x) = k - x$. So that individuals going to the bar positively benefit if attendance is less than or equal to k, but would have done better to stay at home (which we assume has payoff 0) if attendance is greater than k. The El-Farol bar problem is an example of an n-player social parasitism game (Parasitism, chapter 10) and as such we can find the evolutionary stable strategy for bar goers. In this case the evolutionary stable strategy is to attend the bar with probability k/n. Individuals who go to the bar any less than this risk missing a good night out, while those going more often will find it too crowded.

While a probability of visiting the bar of k/n may be evolutionarily stable, whether this equilibrium is reached depends on the mechanisms by which individuals learn whether or not they should attend the bar. We can draw a parallel between the El-Farol bar problem and the model in box 8.A. In particular, we can see the bar as a resource and tables at the bar as resource units and think of the active feedback as individuals in the non-attending population as going to the bar if they hear that a friend went there and found a vacant table. With this interpretation, if bar goers make their decisions independently of each other then bar populations will stabilize. On the other hand, if there is copying and active regulatory feedback then bar populations may over-compensate for previous observations. In particular, if the positive experiences of those attending the bar are communicated to large numbers of non-bar goers (i.e., b in the model in box 8.A is large) then the population of bar goers can oscillate or become chaotic. Individuals who copy others can produce attendance levels that fluctuate wildly around the equilibrium (figure 8.2e, f).

The study of the El-Farol bar problem, and that of the related minority game (Challet & Zhang 1997), has focused on passive regulation. The key question is how individuals that are boundedly rational, being equipped with a small number of strategies and limited memory of previous bar visits, can choose the best bar attendance strategy. On each round of bar attendance each agent adopts the strategy that would have maximized its payoff on previous rounds. When individuals have a small number of strategies to choose between and limited memory of past interactions they do not always converge to the evolutionary stable strategy of attending the bar k/n of the time. As the memory of agents increases, the agents become more efficient and average attendance becomes close to k/n, with only small fluctuations away from this equilibrium (Challet & Zhang 1998; Savit et al. 1999). Interestingly, agents with intermediate memory produce smaller fluctuations in attendance than those with very long memory. This phenomenon occurs because individuals with a long memory can effectively adopt the random, evolutionary stable attendance strategy, while those with an intermediate memory cannot. The conclusion is that agents with limited memory in some cases reduce and in other cases increase a market's volatility, relative to that produced by all-knowing rational agents.

In the El-Farol bar problem there is an advantage to not following the herd, i.e., going to a bar on occasions when others are likely to stay at home. In this case, active regulation or copying others is unlikely to be a desirable strategy. In general, however, and as we saw in chapters 3 through 5, copying can be a good strategy for making decisions in environments where information is limited. In financial markets, there is empirical evidence that financial analysts follow the buy/sell recommendations of their peers (Walter & Weber 2006; Welch 2000), although it is not clear whether this is naive herding or due to correlations in the information used by the analysts (Bernhardt et al. 2006). Copying or herding is the basis of a large number of models of financial markets (Avery & Zemsky 1998; Devenow & Welch 1996; Kirman 1993; LeBaron 2006; Lux 1995; Sornette 2003a). As such, these models are similar to those of preferential attachment (box 2.C) and positive feedback (box 3.A) discussed earlier in this book. The properties that emerge from these models, such as power law distributions and sudden changes in group dynamics in response to small changes in model parameters, also correspond to those observed in real markets.

There is still a great deal of debate about the key mechanisms underlying market stability and fluctuations. However, the "herding" models and other models arising from theoretical physics take us away from the view of the self-regulating noisy market, towards a view of a market that is sometimes stable, sometimes unstable, but always complex (LeBaron 2006; Shiller 2000; Shiller 2003; Sornette 2003b). The wealth of data available on financial markets makes them an ideal system for studying human collective behavior, and they will continue to fascinate scientists for many years to come.

Congestion

Many species of ants form well-defined trails between food and the nest (see chapters 3 and 7). These trails are used simultaneously both by outbound and inbound ants, potentially leading to congestion on the trail and reduced traffic flow. In the leaf cutting ant, *Atta cephalotes*, encounters between ants moving in opposite directions slow their average walking speed by 16% for inbound and 21% for outbound ants (Burd & Aranwela 2003). Given the importance of rapid delivery of food to the colony, we might expect these ant species to evolve mechanisms for efficient flow on trails. Minimization of collisions is also important in human pedestrian traffic. The design of walkways and safety exits involves optimally regulating the flow of humans through their environment (Batty et al. 2003b; Helbing 2001; Helbing et al. 2000; Helbing et al. 2007).

Theoretical models predict that individual ants or humans do not have to adopt particularly sophisticated movement rules in order for an efficient trail organization to emerge. Helbing & Molnár (1995) modeled lane formation in pedestrians, using a self-propelled particle approach (see chapter 5). Each pedestrian particle was equipped with only two rules: walk in a particular direction and avoid collisions with others. At high densities of pedestrians moving down a corridor in opposite directions, collisions are frequent at first. Collisions are reduced for an individual pedestrian if it, through a random sequence of collision avoidances, finds itself behind another individual moving in the same direction. These "traffic lanes" are stable, since leaving the lane will result in increased collisions. Eventually the traffic segregates into unidirectional lanes that provide highly efficient traffic flow (Helbing 2001; Helbing & Molnár 1995). An example of the resulting lane structure is shown in figure 8.3a.

(a)

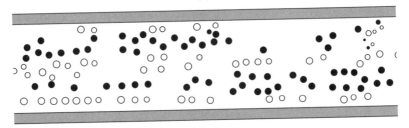

(b)

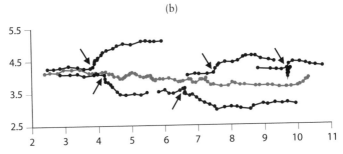

Figure 8.3. (a) Model predictions of pedestrians forming unidirectional lanes of traffic. The black and white circles represent pedestrians moving in opposite directions (reprinted with permission from D. Helbing & P. Molnár, "Social Force Model for Pedestrian Dynamics," May 1995, *Physical Review* E 51, 4282–4286, fig. 2, © the American Physical Society, http://link.aps.org/doi/10.1103/PhysRevE.51.4282. (b) Traffic flow on army ant trails showing the path taken by a nest-bound ant interacting with five outbound ants. Arrows indicate the points at which the ants interacted (reproduced from I. D. Couzin & N. R. Franks, "Self-organized lane formation and optimized traffic flow in army ants," September 2003, *Proceedings of the Royal Society B: Biological Sciences*, 02PB0606.1–02PB0606.8, fig. 3b, © The Royal Society).

Lane formation occurs on the fast moving and densely populated trails of the army ant, *Eciton burchelli*. The traffic separates into three lanes, of which the inner lane consists primarily of nestbound ants and the outer lanes of outbound ants (Couzin & Franks 2003). This three-lane formation, as opposed to the multi-lane formation predicted by Helbing & Molnár's model, results from a larger turning angle when avoiding collisions by outbound ants. These ants then move to the edge of the trail while inbound ants continue down the center (figure 8.3b).

A related finding of Helbing et al.'s model is that, when constrained to move through a narrow door the flow of traffic will oscillate between the two directions (figure 8.4). This temporal organization again arises because those entering the door behind another individual are less constrained than those attempting to enter a door from which individuals

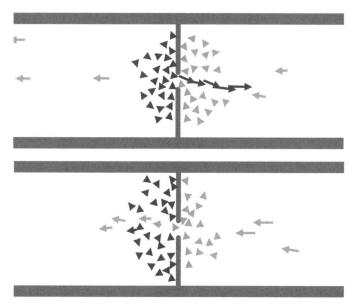

Figure 8.4. Two snapshots of a simulation of passage through a narrow door show oscillations in the build up and flow in opposite directions (reproduced from D. Helbing, P. Molnár, I. J. Farkas, & K. Bolay, "Self-organizing pedestrian movement," *Environment and Planning B: Planning and Design* 28(3) 2001, 361–383, fig. 7, © Pion Limited, London).

are leaving. Dussutour et al. (2005b) tested this prediction by restricting foraging *Lasius niger* ants to a bridge with a width sufficient to allow only two ants to stand side by side. Alternating clusters of inbound and outbound traffic used the bridge. The ants exhibited an additional element of co-operation, over and above that defined in the Helbing model. Ants at the bottleneck gave way to ants already traveling towards them on the bridge, waiting until a gap arose in the flow and the direction switched (Dussutour et al. 2005b). This temporal organization meant that the flow rate on the narrow bridge was equal to that on a bridge more than 3 times as wide.

In ant foraging, the regulatory feedback of avoiding collisions is complemented by positive feedback provided by pheromone recruitment. Pheromone recruitment is a form of active regulation, which increases the number of ants taking a particular route. Interactions with others act as passive regulation that prevents ants taking overcrowded routes. Dussutour et al. (2004) took the pheromone recruitment model presented in box 3.A and extended it by adding a term for overcrowding. In the standard model symmetry-breaking bifurcation occurs, where for sufficiently high flows of ants one of the bridges was used more often than the other. With the inclusion of a term for overcrowding, however, as the width of the bridges decreased a bifurcation occurred whereby at high flow rates the bridges were exploited equally again. Dussutour et al. (2004) tested these predictions with *Lasius niger* ants and confirmed that as bridge width was narrowed the ants switched from using one to two bridges.

Ant trails are good examples of how simple strategies of avoiding collisions can lead to efficient traffic flow at the group level. However, Burd (2006) points out we should not overestimate the importance of flow of ants as a measure of efficiency of these trails. Encounters between in- and outbound ants are important sources for transfer of materials (Anderson & Jadin 2001) and, potentially, information (Burd & Aranwela 2003).

Preliminary experiments on human pedestrian traffic and "field" observations appear to confirm many of the predictions of Helbing's models (Helbing et al. 2005; Kretz et al. 2006a, 2006b, 2006c). One of the most interesting outcomes of these experimental studies is that the sum of the flows in both directions of bidirectional traffic is higher than the flow of similar densities of pedestrians moving in a single direction (Helbing et al. 2005; Kretz et al. 2006a). The challenge now is to disentangle sociological explanations for phenomena, e.g., people walk more rapidly when approaching individuals moving in the opposite direction, from explanations based on collision avoidance and interaction radii (Moussaïd et al. 2009). An area where this work has had high impact and useful consequences is control at events such as football matches or religious gatherings that attract large crowds (Batty et al. 2003a, 2003b; Helbing et al. 2007).

Segregation and Self-sorting

Humans and other animals often regulate who they interact with. Schelling proposed a series of models aimed at disentangling the mechanisms by which individuals become segregated (Schelling 1969, 1971, 1978). He uses the rather provocative example of racial segregation to illustrate his model, but his approach illuminates how individuals become sorted in everything from age and income to their hobbies and consumer preferences. One version of this model is described in box 8.B. The model assumes a city neighborhood made up of populations of both blacks and whites. Each individual has their own tolerance level for the ratio of whites to blacks. The tolerance level across individuals can be thought of as a distribution, with some individuals more tolerant than others. Schelling assumed that individuals would leave the neighborhood if the race ratio were out of the limits of their tolerance and move in if the

Box 8.B Segregation Model

Schelling (1969) proposed a model of racial dynamics within a city neighborhood inhabited by people who are either black or white. He assumed that white people are willing to live with a particular ratio of blacks to whites, and this ratio is different among different individuals. An example of a cumulative distribution for the white's "tolerable" race ratio is given in figure 8.5a. Under this distribution all whites tolerate being in a neighborhood that contains only whites, half of them would happily be in a neighborhood with a 1:1 equal ratio, and a small number of them would tolerate an almost 1:2 whites to blacks, leaving them in a minority. This tolerance distribution thus reflects differences within the population of whites for tolerating blacks in their neighborhood.

We write W, respectively B, as the number of whites, resp. blacks, living in the neighborhood out of a potential population of N_W, resp. N_B, whites, resp. blacks, who could choose to live in the neighborhood. Thus W/N_W is the proportion of whites and B/N_B is the corresponding proportion of blacks living in the neighborhood (note that the proportions are calculated relative to the potential inhabitants of each race rather than the proportions of blacks vs. whites living in the neighborhood). We can now write an expression, which is shown in figure 8.5a, for the proportion of whites tolerating a neighborhood: $P_W = 1 - B/rW$, where 1:r is the ratio above which no white will move into a neighborhood.

There are various ways in which the tolerance distribution can influence the behavior of individuals and thus the dynamics of the racial mix of a neighborhood. Schelling (1969) assumed that if the actual proportion of whites living in the neighborhood is greater than the proportion that would tolerate the current ratio, i.e., $W/N_W > P_W$, then whites would start to leave the neighborhood, starting with the least tolerant. Similarly, if the actual proportion of whites living in the neighborhood is less than the proportion that would tolerate the current ratio, i.e., $W/N_W < P_W$ then whites would move into the neighborhood. Rearranging terms in these equations we see that whites will move out of a neighborhood if $B > rW(1 - W/N_W)$ and into a neighborhood if $B < rW(1 - W/N_W)$. The curve $B = rW(1 - W/N_W)$, plotted in figure 8.5b, gives

(Box 8.B continued on next page)

191

the equilibrium at which whites will no longer move in or out of a neighborhood.

If only whites make movement decisions on the basis of the racial makeup of a neighborhood then the neighborhood's composition will eventually reflect the tolerance of the white population for mixed race neighborhoods. However, if blacks simultaneously make movement decisions based on race ratio then this outcome changes. Figure 8.5c shows the effect of whites and blacks simultaneously making movement decisions when both races have identical tolerance distributions for each other and $r = 2$. When the neighborhood at first contains a small number of whites and blacks with a near 1:1 ratio then both blacks and whites will move into the neighborhood. However, if the ratio is biased slightly in one direction, say with a small majority of blacks, then as the population of the neighborhood increases whites will start to move out and more blacks will move in. As this process continues, the ratio will change to a larger majority of blacks and even more whites will move out and blacks move in, distorting the ratio still further. The stable ratio of blacks to whites is then 1:0. A similar argument applies if the initial majority are whites, with a stable ratio of 0:1. Neighborhoods with a small black majority become all black, and those with a small white majority become all white. This situation changes when both races show a higher tolerance. Figure 8.5d shows that when $r = 4$ the 1:1 ratio is locally stable, although the 1:0 and 0:1 ratios also remain locally stable. Which of these equilibrium ratios occurs depends upon the initial ratio of blacks to whites.

ratio were within their tolerance limit. Through this process a pattern of segregation emerges for the racial make-up of the neighborhood.

The striking aspect of Schelling's model is that even a tolerance distribution that might initially appear relatively tolerant (in the wider meaning of the word) can lead to segregation. In the example in figure 8.5a, 50% of white individuals would accept living in a neighborhood consisting of half whites and half blacks. The model predicts, however, that the only stable ratios of blacks to whites are single race neighborhoods (figure 8.5c). Figure 8.5d shows that when 75% of individuals accept a 1:1 ratio then this ratio is stable, but the all white and all black neighborhoods still remain stable if the initial populations lie nearby. The message of this model is clear: the fact that people tolerate a degree of racial mixing does

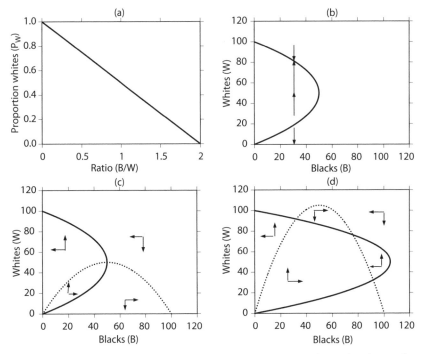

Figure 8.5. Schelling's neighborhood segregation model. (a) Cumulative distribution for white's "tolerable" race ratio. In this case $P_W = 1 - B/rW$, with $r = 2$. (b) Equilibrium level at which whites neither move into nor out of a neighborhood. The arrows indicate for a particular population size of whites, whether the white population will increase (i.e., individuals move in) or decrease (i.e., individuals move out). (c) Equilibrium level at which whites neither move into nor out of a neighborhood, plotted together with the equilibrium level at which blacks neither move into nor out of a neighborhood. The arrows show the direction in which the population changes. The population stabilizes at either a 1:0 or 0:1 race ratio; $r = 2$ for both whites and blacks. (d) The same as (c), but for $r = 4$. In this case, the ratio 1:1 becomes stable.

not imply that segregation is avoided. Likewise, the existence of complete segregation does not imply complete intolerance on the part of the people living in the segregated neighborhoods. Although individuals may think they are regulating their relocation decisions in a way that will generate an integrated society, the outcome can be highly segregated.

Real tolerance distributions are not as simple as those assumed in Schelling's model and they are known to differ between whites and blacks. Numerous studies based on both questionnaires and measurements of peoples' behavior show that in the USA during the second half of the 20th century very few whites would tolerate living in an all black neighborhood and a minority would tolerate living with more than 50%

blacks. A sizeable minority of blacks would tolerate living in an all white neighborhood and nearly all would tolerate a 50% white neighborhood (Bruch & Mare 2006). Furthermore, in a study by Farley (1978), 79% of whites in Detroit said they would be very comfortable in neighborhoods that were not more than 20% black.

Clark (1991) used the then available survey data to make empirical preference distributions to test Schelling's model. Analysis of four American cities showed results qualitatively similar to that of Schelling's original model: mixed race equilibriums are unstable, while all black and (nearly) all white equilibriums are stable. One difference between the empirical data and the original model is that a small number of blacks can stably inhabit a predominately white neighborhood. Another difference is that the region where both races move into an area is smaller than that predicted in figure 8.5c. The conditions under which both whites and blacks will move into an area are very limited, and ultimately Clark (1991) predicts racial segregation.

The question that individuals were asked in Clark's study was which ratio of blacks to whites they *preferred*. In his original formulation, however, Schelling uses the term *tolerance* to denote a cut off point between wanting to leave and wanting to move into a particular neighborhood. For example, a white individual who would ideally live in and actively seek out a neighborhood with 50% blacks, and is neutral to living with up to 75% blacks but prefers not to live with a higher ratio than this is said to have a tolerance of 75%. Clark's notion of preference does not account for such neutral tolerance. Studies by Farley (1978) and Farley et al. (1994) presented interview subjects with a sequence of cards on each of which was drawn a simple representation of neighborhoods consisting of 15 houses, a proportion of which were coloured black, to represent black occupants, while the others were white. They asked whether the subjects would move into the area represented by each card, increasing the number of non-like neighborhoods on each of the cards shown. Figure 8.6 shows the outcome of Schelling's model given the "moving in" distributions established in these studies. In this case, the mixed equilibrium would be stable, although only marginally so, under the results of the questionnaire in 1992 (figure 8.6a) but not in 1978 (figure 8.6b). In reality, the current racial makeup of the city of Detroit, whose residents were the subject of Farley's study, is not consistent with a mixed equilibrium. The 2001 census showed that central Detroit has over 80% black and less than 13% white inhabitants, with many whites having moved to the suburbs.

When moving away from illustrative models of segregation and toward applications to real world data, it is important to remember the difference between correlation and causation. There is little doubt that

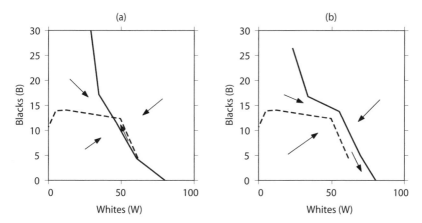

Figure 8.6. Tolerance curves from questionnaires collected from Detroit residents from (a) 1978 and (b) 1992 (Farley 1978; Farley & Frey 1994). Equilibrium level at which whites neither move into nor out of a neighborhood, plotted together with the equilibrium level at which blacks neither move into nor out of a neighborhood. The arrows show the direction in which the population changes.

when asked about a neighborhood with a particular racial makeup, interview subjects base their opinion not simply on their like or dislike for a particular race but on their idea of: the quality and prices of housing and schools; crime levels; and social problems in areas that have that racial structure at the time of the interview. Disentangling whether it is race per se, or correlated variables, or even perception of correlated variables that determines people's relocation decisions is a challenging and important problem, and one about which Schelling's model says very little (Charles 2003). It is also important to bear in mind the mathematical limitations of Schelling's model. Most importantly, the model assumes a fixed population of blacks and whites that choose whether or not to live in a particular area. It does not say what happens to individuals who choose to live elsewhere or, due to high house prices in certain neighborhoods, have no choice about where to live. These limitations aside, Shelling's model powerfully illustrates that weak preferences can generate strong segregation, making it all the more difficult to counteract.

The relationship between preference at the level of the individual and aggregation or segregation at the level of the group is complex. For example, Rivault et al. (1998) found that different strains of cockroaches preferred the odor of their own strain. However, in experiments in which cockroaches of the two strains were put together in an arena with two shelters, they aggregated all under the same shelter (Amé et al. 2004). In this case, aggregation instead of segregation occurred because, although each strain preferred its own odor, it was weakly attracted to the odor of the other strain. This weak attraction was amplified when the strains were put together. Millor et al. (2006) showed (using a model that is a two strain extension of that in box 3.A) that whether segregation or aggregation occurs between two strains that are weakly attracted to each other depends upon the size of the groups and the relative between- and within-strain attraction. Large groups of individuals with weak between-strain attraction will segregate, while smaller groups of individuals with strains that are more strongly attracted to each other will aggregate.

The theoretical results are partially supported by experiments that placed in the same arena equal numbers of two different species of cockroach: *Periplaneta fuliginosa*, which is strongly attracted only to the odor of conspecifics; and *Periplaneta Americana*, which is weakly attracted to the odor of conspecifics as well as to that of *P. fuliginosa* (Leoncini & Rivault 2005). In these experiments segregation was a more common outcome than aggregation. However, smaller groups of ten cockroaches aggregated more often (in 38% of trials) than larger groups of twenty cockroaches, (which aggregated in 19% of trials). The most interesting conclusion of these theoretical and empirical studies is that aggregation and segregation both arise from the same set of individual rules, with initial distribution, group size, and between-group preference playing an important role in outcome. As with segregation in human societies, we should not conclude that just because two groups segregate that they are necessarily intolerant of one another.

Active regulation on the basis of particular characteristics is seen within fish shoals. Croft et al. (2003) showed that when shoals of guppies split, they actively segregated in terms of body length. They were, however, also sorted on the basis of their response to their environment. Larger fish were more often found further from the surface of the water. These observations have interesting consequences for the social structures of shoals (Croft et al. 2006, 2005). Segregation leads to particular forms of social networks, which in turn determines how information flows through a population (Newman 2003; Watts & Strogatz 1998).

An Invisible Hand?

Adam Smith's invisible hand, which guides the economy to equilibrium, remains a powerful metaphor for thinking about regulation. Many economical systems are close to equilibrium and function very efficiently without centralized control. One of the most remarkable examples of an invisible hand we have seen in this chapter is the lane formation in ants and humans. Simply by avoiding collisions, individuals self-organize in lanes that allow for efficient flow. Unfortunately, the invisible hand is not always as steady as we might hope. We have seen how active regulation can lead to over-compensation and chaos and how even a small tendency to prefer associations with like individuals can lead to strong forms of segregation. It is amazing how stable the social world is to perturbations, but it is also worth remembering how easily it can spiral out of control.

— Chapter 9 —

Complicated Interactions

Throughout this book I have emphasized how simple rules followed by individual animals and humans can produce surprisingly complex patterns. It is this observation, combined with the idea that we can use mathematical models to predict these patterns, upon which the idea of self-organization is founded (Camazine et al. 2001; Nicolis & Prigogine 1977). Indeed, it is common to hear these "complex systems" contrasted with "complicated systems." The former term is associated with systems in which complexity emerges from simple interactions, while the latter is associated with systems where large numbers of different components, each with its own particular role, interact to produce an output. The contrast is best illustrated by examples from physics. An example of a physicist's complex system is a sandpile. When grains of sand are dropped from above onto a particular position, a pile builds up and sand moves down the outside of the pile. The movement of sand on the outside of the pile is difficult to predict and occurs on scales ranging from small local toppling to large avalanches. Removing one or two grains of sand will not change this overall pattern. A car or an airplane, on the other hand, can be thought of as complicated. It consists of lots of parts that are carefully put together to drive from A to B. Removing certain components can completely change the car's capability of completing its journey.

Are animal groups sandpiles or cars? Up until now, I have emphasized the former analogy. However it is often the second analogy that is more appropriate when studying animal interactions. For example, individual honeybees are known to use at least 17 different communication signals, the most famous of which is the waggle dance, and adjust their behavior in response to at least 34 different cues (Seeley 1998). The bees take different behavioral roles at different times during their life. Furthermore, there are certain components, such as the queen, which are essential to the smooth functioning of the colony.

In general, when building mathematical models the question of whether a system is complex or complicated is not a particularly useful one to ask. Rather, the question is whether there is a level of description at which we can formulate a model that answers our questions about a system's behavior (Goldenfeld & Kadanoff 1999). As the preceding chapters have demonstrated there often exists such a level and mathematical models do help our understanding of collective animal behavior. There is however no reason to believe that this level of description can be identified in all cases. For example, although in chapter 3 I showed how a model predicts how honeybees and ants balance their foraging across feeders of different quality, this does not answer the larger question of how the colony regulates its overall growth. A successful colony must balance its requirements for foraging with other tasks such as building and nursing brood (Gordon 1996). Even if we concentrate only on nectar foraging we see that honeybees exhibit at least seven different behavioral states, e.g., scout, recruit, inspector, etc. (Biesmeijer & de Vries 2001), and exhibit a range of signals about the location and availability of food (Seeley 1995). We can use simple models to focus on understanding particular parts of this organization, but these do not necessarily provide a level of description that explains how the colony functions as a whole.

In this chapter I look at models that attempt to capture more fully the detailed interactions within insect societies, in particular. As hinted at in the preceding paragraphs, one of the best studied systems in this context is the foraging of honeybees. Another well studied system is the emigration of Temnothorax ants, and it is this system on which much of this chapter will focus. Here, I introduce the use of state- and agent-based models, using foraging and emigration of social insects as case studies around which the various techniques are discussed.

Social Insect Foraging

Behavioral State Modeling

Seeley (1995) builds an understanding of honeybee organization by identifying how individuals moved between *behavioral states* in response to signals they received from other bees and cues they received from local observations of the state of the colony. This approach lends itself naturally to some form of agent-, individual-, or state-based modeling. If we can write down the behavioral states that an individual or agent can exhibit, and determine the rate at which they make transitions between these states, then we can write down a model of each honeybee's behavior. The agents interact with each other by making these transition rates change as a function of the number of individuals in other behavioral states.

One of the first examples of this approach, and one that we came across in chapter 3, is Camazine and Sneyd's model of honeybee foraging (Camazine & Sneyd 1991). The behavioral states in this model included

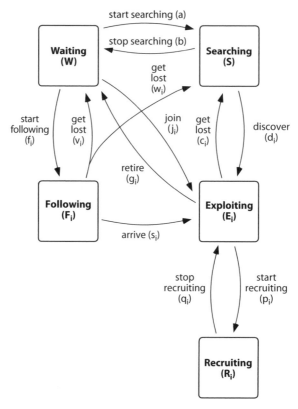

Figure 9.1. Flow diagram for behavioral state variables for Sumpter & Pratt's framework (box 9.A). Boxes represent behavioral states, while lines connecting states indicate rate of flow of workers between states. Arrows indicate direction in which individuals change states.

searching for a food source, performing dances, and following dances in the hive. The transitions between behavioral states depended on the states of other individuals. For example, the transition from following dances to searching for a particular source depended on the number of individuals dancing for that source.

Following from the Camazine and Sneyd model, Sumpter and Pratt (2003) proposed a general framework for modeling social insect foraging in terms of differential equations, based on transitions between behavioral states. This framework is described in box 9.A. The five basic behavioral states are waiting, searching, following, exploiting, and recruiting (figure 9.1). For nectar foraging of honeybees: waiting corresponds to waiting on the dance floor in order to follow a dance; following corresponds to

Box 9.A State-based Models of Foraging

Sumpter & Pratt's (2003) framework defines five different behavioral states associated with foraging. Colonies are assumed to have access to n food sources (e.g., patches of flowers or sugar feeders). Each state has an associated variable, indexed by source where appropriate, representing the number of individuals in that state. The states (and corresponding variables) are:

1. Waiting (W) Waiting at the nest and available to start foraging. Examples include honeybees waiting on the dance floor to follow recruitment dances, or ants waiting near the nest entrance to be led to a food source.
2. Searching (S) Searching for food sources.
3. Exploiting (E_i) Exploiting food source i. Workers in this state do not directly recruit nest mates, although they may leave signals, such as pheromone trails, that increase the likelihood of other foragers finding the source.
4. Recruiting (R_i) Attempting to recruit nest mates to food source i. Recruitment in this sense involves actively leading one or more workers, or directly communicating to nest mates the location of a food source, rather than leaving chemical signals in the environment.
5. Following (F_i) Attempting to follow recruiters to food source i. This encompasses not only literal following of recruiters, but also independent search for a source advertised by a dance or other signal.

Figure 9.1 shows how individuals change between states. For example, an individual becomes a follower from the state of waiting at the nest, and from following it can either get lost or arrive at its target.

In order to model these states in terms of differential equations we must specify the rates at which an individual changes from one state to another. For example, assume that λ is the probability that in a small time interval (dt) a waiting individual starts to follow dances. W is the number of waiting individuals. We then make the mean-field approximation that the rate at which a population of

(Box 9.A continued on next page)

waiting individuals is converted into dance-following individuals is equal to λW, i.e.,

$$\frac{dW}{dt} = -\lambda W.$$

This approximation ignores any random variation or differences between individuals, and since λW is not an integer, it also ignores the fact that bees come in distinct entities. Despite these limitations such approximations work well provided the number of individuals in each behavioral state is relatively large (in practice this is more than 5 to 10 individuals). Thus, although differential equation models are ultimately written in terms of populations, the equations are initially derived from individual behavior.

Behavioral transition rates usually depend on the number of individuals in another state. For example, the probability that dance-following honeybees start looking for feeder 1 is proportional to the number of bees dancing for that feeder, i.e.,

$$\frac{R_1}{R_1 + R_2 + K_0},$$

where K_0 is constant and R_1 and R_2 are the number of bees dancing for feeders 1 and 2, respectively. We can express the rate of change of the number of following bees as

$$\frac{dF_1}{dt} = \lambda \frac{R_1}{R_1 + R_2 + K_0} W - \theta_1 F_1,$$

where λ and θ_1 are constants determining the rate per individual bee of starting to follow dances and getting lost, respectively. Similar equations can be written down for each behavioral state giving a system of differential equations modeling how individuals change between behaviors. Based on the earlier model of Camazine & Sneyd (1991), Sumpter & Pratt proposed the following differential equation model for honeybee foraging

$$\frac{dW}{dt} = \sigma_1 E_1 + \sigma_2 E_2 + \theta_1 F_1 + \theta_2 F_2 - \lambda W + \gamma S$$

$$\frac{dF_1}{dt} = \lambda \frac{R_1}{R_1 + R_2 + K_0} W - \theta_1 F_1 - \phi_1 F_1$$

$$\frac{dF_2}{dt} = \lambda \frac{R_2}{R_1 + R_2 + K_0} W - \theta_2 F_2 - \phi_2 F_1$$

$$\frac{dS}{dt} = \lambda \frac{K_0}{R_1 + R_2 + K_0} W - (\alpha_1 + \alpha_2 + \gamma) S$$

$$\frac{dE_1}{dt} = \phi_1 F_1 + \alpha_1 S - \sigma_1 E_1 - (\rho_1 E_1 - \delta_1 R_1)$$

$$\frac{dE_2}{dt} = \phi_2 F_2 + \alpha_2 S - \sigma_2 E_2 - (\rho_2 E_2 - \delta_2 R_2)$$

$$\frac{dR_1}{dt} = \rho_1 E_1 - \delta_1 R_1$$

$$\frac{dR_2}{dt} = \rho_2 E_2 - \delta_2 R_2 \qquad (9.A.1)$$

The various parameters in this model have been measured directly from experimental data and a simulation of this model for the experimental setup of Seeley et al. (1991) is given in figure 9.2.

searching for nectar source advertized by a dance; searching corresponds to scouting for food without first following a dance; exploiting involves flying backwards and forwards to a known food source; and recruiting involves performing recruitment dances. Similar interpretations can be made of the foraging states of ants, but with pheromone trails providing direct recruitment from waiting to exploiting instead of the indirect recruitment provided by the dance language.

In the framework in box 9.A, the behavioral states, the transitions between the states and associated rate parameters can be determined by observations of individuals. The differential equation model is then written in terms of the number of individuals in each of the states. The assumption underlying the change from individual to population description is known as the law of mass action or mean-field approximation. The basic idea of this assumption is that when considering large numbers of individuals in the same state we do not need to consider every individual, but instead consider simply the rate at which populations of individuals switch between states (see box 9.A).

Despite the differential equation model being an approximation of individual behavior, it often works reasonably well in predicting colony level behavior. Figure 9.2 shows a numerical solution of the differential equation model for honeybee foraging presented in box 9.A for an experiment performed by Seeley et al. (1991). Figure 9.3a shows the outcome

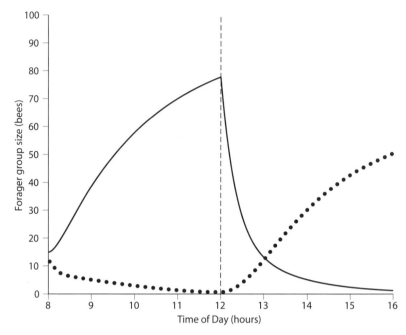

Figure 9.2. Numerical simulation of equations 9.A.1 showing total number of recruiting and dancing bees for the two sites. Solid line gives number at south feeder $(E_1 + R_1)$ while dotted line is the north feeder $(E_2 + R_2)$. As in fig. 9.3, the simulation begins with $E_1(0) = 15$, $E_2(0) = 12$, and $W(0) = 125 - 15 - 12 = 98$, and at time 12 the quality of the feeders is swapped. Simulation parameter values can be found in Sumpter & Pratt (2003).

of Seeley et al.'s experiment. The model reproduces these experimental results reasonably accurately, although it underestimates the rate at which the bees switch feeders when the quality of the feeders is switched.

Complicated Individuals

Seeley describes the foraging of a honeybee colony as "an ensemble of largely independent individuals that rather infrequently exchange information with one another" (Seeley 1995). Seeley's emphasis is on conservation of communication. Rather than simple units using mass communication to form a collective solution, complicated individuals use the minimum of communication necessary to co-ordinate their work (Seeley 2002). An individual honeybee or other foraging social insect is more complicated than implied by the framework in box 9.A. Biesmeijer & de Vries (2001) propose that the behavioral states of honeybees should include different categories for novice forager, scout, recruit, employed forager, unemployed experienced forager, inspector, and reactivated forager. Their proposal is

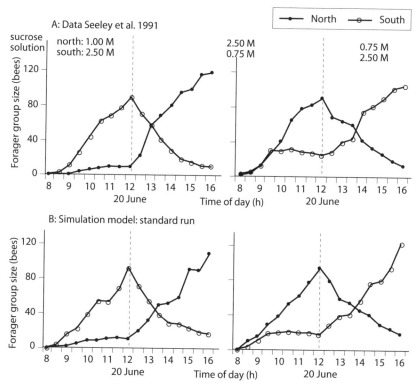

Figure 9.3. (a) Number of honeybee foragers visiting each of two feeders (north and south) recorded during 30-min intervals in the experiment of Seeley et al. (1991). In the experiment both feeders were 400 m from the hive and their quality, determined by the molarity (M) of the sugar solution, was changed at the beginning of each day and at noon. (b) Simulation outcome of de Vries and Biesmeijer's individual-based model of these experiments. (Reproduced from H. de Vries & J. C. Biesmeijer, "Modelling collective foraging by means of individual behaviour rules in honey-bees, 1998, *Behavioral Ecology and Sociobiology* 44:2, 109–124, fig. 3, © Springer-Verlag.)

based on the studies of von Frisch (1967), Lindauer (1952), through to Seeley (1983, 1995) where honeybees are shown to use a combination of personal information of where food is located and social information gained through following dances.

A similar individual complexity underlies ant foraging (Detrain & Deneubourg 2006; Gordon 2007). For example, harvester ants do not usually rely on pheromone trails to co-ordinate foraging, but foragers resting within the nest are activated by other foragers and by patrolling ants (Gordon 2002; Gordon et al. 2008; Greene & Gordon 2007b). Individual foragers have a memory of their own foraging patch and activation by other ants causes them to visit their own patch rather than to blindly follow the activating ants. The role of communication in this case is for ants outside the nest to indicate to ants within the nest that foraging conditions are generally good (Gordon 2007).

The advantage of state-based models is that there is no limit to the complexity that can be incorporated into them. However, increasing the number of behavioral states means decreasing the number of individuals in any particular state at any particular time. As such, the mean-field assumption underlying differential equation models can no longer hold. If only a few individuals are in a particular state, then we cannot assume that the transition between the states can be approximated as a transition rate of populations. Furthermore, if only one or sometimes no individuals are in a particular state, then our differential equation model will represent fractional individuals.

Added complications on the level of individual behavior can be incorporated into agent-based or individual-based models that preserve individual units and the stochastic nature of interactions. De Vries & Biesmeijer (1998, 2002) developed an individual-based model of honeybee foraging where each bee was characterized by its position, speed, direction, visual perceptions, its memory of the position and profitability of food, as well as internal motivation for homing, foraging, and abandoning a food source. By fitting their model to Seeley et al.'s (1991) experiments they showed that previous foraging experience, and not only recently acquired dance information, were required to reproduce the experimental results (de Vries & Biesmeijer 1998). In doing so they identified a model that was sufficient to explain current experimental observations. An example simulation outcome compared to data from the experiment is given in figure 9.3b. Here the match between experiment and data is better than in the original differential equation model (figure 9.2).

De Vries & Biesmeijer's model is a useful tool for investigating hypotheses about individual honeybee behavior. Several other authors have provided their own individual-based models of honeybee foraging (Beekman & Bin Lew 2008; Dornhaus et al. 2006) and other aspects of honeybee organization (Schmickl & Crailsheim 2008a, 2008b). This approach is, however, limited in several respects. Firstly, these models have a large number of parameters, many of which could not be estimated from independent datasets. Secondly, while models are usually sufficient to explain the data, it is difficult to argue that a particular model is necessary. In other words, there may exist other models that fit the data equally well. Both these limitations of individual-based models are a consequence of insufficient data and could be resolved by more experiments. I will return to this question in more detail in later sections of this chapter in relation to ant and honeybee migration.

Complicated Signals

Not only do individuals have large numbers of behavioral states, but their behavior also is influenced by a diversity of cues and signals. For example, foraging bees are influenced in their decision whether to dance by both their own assessment of the quality of food they carry as well as the time it takes them to unload the nectar they bring into the hive (Seeley 1992, 1995). Ant foraging is also more complicated than implied in chapter 3. Individual ants use their own information, as well as that gained through interactions with others, both to locate food and to decide whether to recruit to it. For example, a *Lasius niger* ant's decision to leave a pheromone trail to a food source depends on how easily it reaches its desired volume of food (Mailleux et al. 2000, 2003b), its level of starvation (Mailleux et al. 2006) and the nutritional needs of the colony (Portha et al. 2004).

Ants use a variety of pheromones to mark the path to food discoveries (Wyatt 2003). For example, some *Myrmica* species use pheromone from different glands depending upon the type of food they locate (Cammaerts & Cammaerts 1980). Pheromone with stronger recruiting properties is laid to prey that are hard to move, thus recruiting other workers to help with transportation.

Combination of different types of pheromones with different lifetimes may allow ants to "remember" routes to sites that were previously rewarding and may become rewarding again in the near future. Pharaoh's ants provide a good example of an ant that leaves multiple pheromone signals. Jackson et al. (2006) showed that these ants leave a pheromone trail that can be detected up to 2 days after it is laid. The ants deposit pheromone even in the absence of food (Fourcassie & Deneubourg 1994). However, Jeanson et al. (2003) established that the pheromone deposited directly after a food discovery evaporates in less than 25 minutes. Jackson and Chaline (2007) report that the intensity of trail laying, in terms of the degree of continuity of the markings made, changes only slightly between ants returning from a rewarding food source and those exploring. These experiments did not investigate the chemical composition of the trails. Although Jackson and Chaline are cautious about concluding that different chemicals are used for marking during exploration and exploitation, the existence of distinct "explore" and "exploit" pheromones remains the most plausible explanation of the rapid exploitation of newly discovered food (Beekman et al., 2001; Sumpter & Beekman, 2003) and the rapid abandonment of trails that no longer lead to food (Jeanson et al. 2003). These pheromones are possibly complemented by a volatile negative pheromone that serves as a "no entry" signal when food is not found at the extremity of a path (Robinson et al. 2005).

Much of the work on understanding foraging trails is based on behavioral analysis, and less is known about how specific chemical components within these trails act in different circumstances. Chemical communication is also seen in, for example, dance communication in honeybees (Thom et al. 2007). An interesting research challenge is to link together specific chemicals found in communication with observed behaviors.

Combining Complex and Complicated Behavior

By their nature, the complex, self-organized patterns seen in ant trails, nest structures, and bird flocks require large numbers of individuals in order to generate them. Indeed, there is evidence that ant species that typically live in large colonies are more likely to use pheromone trails for communicating the presence of food (Beckers et al. 1989). However, there is little evidence from between species comparison that individual complexity decreases with increased colony size (Anderson & McShea 2001a). Honeybees are just one example of a species with both large colony sizes and individuals that exhibit a complicated array of communication signals and behavioral states.

Some of the most interesting questions in understanding the organization of insect societies involve the interaction of self-organizing patterns with the behavior of individual workers (Detrain & Deneubourg 2006). For example, Beekman et al. (2001) showed that pheromone trails emerge only when the ants in a foraging arena reach a critical density (see chapter 3). Other studies have shown that the foraging behavior of individuals changes with the number of ants in the colony (Mailleux et al. 2003a). We can speculate then that the ants may decide whether to leave pheromone or not based on whether they are sufficient in number to utilize trails. The collective pattern changes individual behavior and individual behavior generates the collective pattern.

While complicated behavioral states, multiple signals, and between-individual variation are all important issues when building models of social insect foraging, these have not been the focus of a great deal of theoretical work. Indeed, the only current solution to modeling these phenomena is to incorporate all the relevant variables and parameters into a simulation model and compare the simulation model to data. It is this approach I now investigate further with regard to another aspect of social insect organization, namely emigration.

Emigration of *Temnothorax* Ants

A large number of studies of social insect foraging in the 1990s were followed at the turn of the century by an increased interest in how social insects migrate to a new nest. In part, this switch of interest from foraging to emigration was due to greater experimental tractability of the latter. Emigrations have a clear beginning and an end. The start can be induced by the destruction of a colony's old nest and the end is marked by a move into their new nest. This has increased the amount of data with which to parameterize individual-based models, allowing these models to be better verified. The following sections describe *Temnothorax* emigration as a model system in the study of complicated state-based behavior. This is a system for which we have been able to clearly identify behavioral states and measure parameter values.

Most studied are the ant emigrations of genus *Temnothorax* (Mallon et al. 2001; Möglich 1978; Pratt et al. 2002). The basic steps of this emigration are given in chapter 4, but I summarize them again here. Each ant begins in an exploration phase during which she searches for nest sites. Once she finds a site she enters an assessment phase, carrying out an independent evaluation of the site, the length of the evaluation being inversely proportional to the quality of the site (Mallon et al. 2001). Once she has accepted the site she enters a canvassing phase, whereby she leads tandem runs, in which a single follower is slowly led from the old nest to the new site. These recruited ants then in turn make their own independent assessments of the nest. Once the nest population has reached a quorum threshold the ant enters a committed phase, rapidly transporting passive adults and brood items (Pratt et al. 2002).

Tandem run recruitment has elements of the simple positive feedback seen in the pheromone trails of *Lasius* ants and the aggregation of cockroaches. Temnothorax emigration is however more than just positive feedback. The four phase decision-making process, the use of a quorum threshold to decide whether to perform a tandem run or a transport (Pratt 2005b; Pratt et al. 2002), and the fact that some ants find both nests and choose the superior one (Mallon et al. 2001), all point toward a more complex migration than seen in cockroach aggregation, for example. We could say that *Temnothorax* ants combine elements of self-organization, whereby a global solution to the problem of finding a new nest emerges from the interactions of multiple ants, with a sophisticated behavioral algorithm, and a process whereby individual ants continually monitor the progress of the emigration and change their behavior accordingly.

The detailed experimental understanding of *Temnothorax* migration has made it possible to determine the ants' behavioral states and the factors influencing transitions between these states. Pratt & Sumpter have systematically refined this behavioral algorithm as new experimental data has become available (Pratt 2005a; Pratt et al. 2002, 2005; Pratt & Sumpter 2006). Figure 9.4 gives a flow diagram for the behavioral states and how the ants transition between them, based on the emigration of

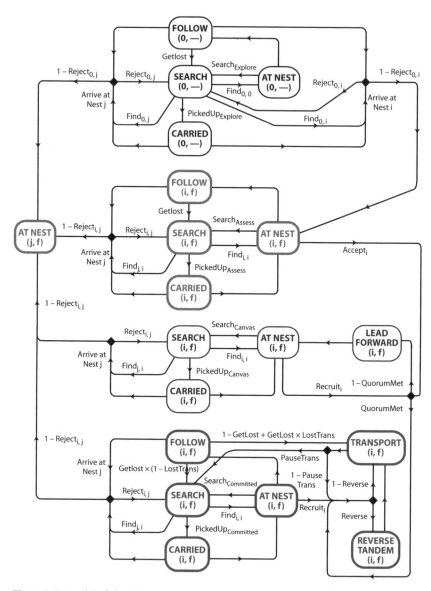

Figure 9.4. Model of the behavior of active ants responsible for organizing emigrations of *Temnothorax albipennis*. Boxes represent behavioral states and arrows represent transitions between them. The four major groups of behavioral states are organized into four groups of boxes. From top to bottom the boxes correspond to exploration, assessment, canvassing, and commitment. The first subscript i in each state identifies the nest that the ant is currently assessing or recruiting to. The second subscript f identifies the nest from which the ant recruits (either the old nest or a rejected new site to which nest mates have been brought by other ants). Figure reproduced from Pratt et al. (2005).

Temnothorax albipennis (Pratt & Sumpter 2006; Pratt et al. 2005). This flow diagram gives a more precise description of the emigration stages described in the previous paragraphs. By individually marking all the ants in the colony and establishing how long and under what conditions they transitioned between states we were able to measure the model parameters (Pratt 2005a; Pratt et al. 2005).

A key purpose of this model is to establish whether our understanding of how the emigration proceeds is correct. If the behavioral algorithm is specified correctly then the output of simulations of the model should be similar to the outcome of experiments. There are a number of ways of making this comparison. Firstly, we can visually compare the sequence of actions performed by the real ants and those in the simulation. An example of such a comparison is given in figure 9.5. Such comparisons are qualitative since both are a single instantiation of the model and the experiment. We can make a series of these comparisons in order to gain insight into differences between model and reality. This approach is often very useful since it quickly reveals differences in the behavior of the real and simulated ants.

A second way to validate the model is to compare distributions of colony level measurements in the model and in the experiment. Using such measurements, the data was shown to not statistically differ from the prediction of the model, i.e., the experimental outcome lay within a confidence interval generated by repeated runs of the model (Pratt et al. 2005). The final way in which the model was validated was to test it against an independent data set. The model was fitted using parameters for single nest emigrations and then tested against the outcome of emigrations to two nests. Again a confidence interval of outcomes was constructed and compared to the actual emigrations.

What can be hidden in the final presentation of a detailed model are all the alternative models that could be rejected by comparison to data. For example, early versions of the Temnothorax model assumed that the time taken for each individual to perform a transport and tandem run was constant, but when this model failed to match the data we realized that individual ants improved their route with repeated journeys between old and new nest sites. As a result we incorporated a travel time that decreases with number of completed journeys and the model provided a better match. The model also revealed that we were not required to assume different parameter values for different individuals within the colony to match the experimental data. This was a particularly striking result since the division of labor through the emigration is suggestive of some ants being more active than others. In particular, the distribution of the number of transports and tandem runs performed by colony members is highly skewed, with some individuals much more involved in the

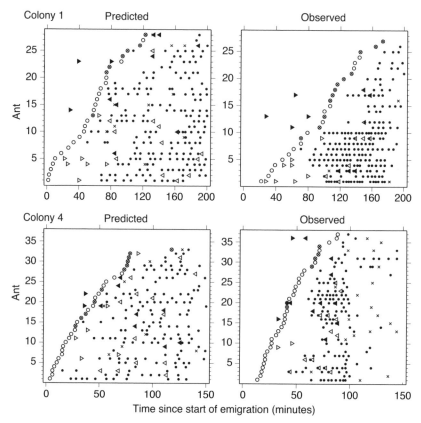

Figure 9.5. Behavioral sequences of active ants, predicted by the model and observed in single-nest emigrations by colonies 1 and 4. Within each panel, each row shows the acts of a single ant. Symbol key— ○: initial entry into the new nest; ▷: leading a tandem run towards the new nest; ▶: following a tandem run towards the new nest; ◁: leading a reverse tandem run; ◀: following a reverse tandem run; ●: transporting a nest mate or brood item to the new site; X: being transported into the new site. Figure reproduced from Pratt et al. (2005).

emigration than others. Our model showed that this division of labor could arise simply as a result of some ants finding and learning the route to the new nest before the others.

The model in figure 9.4 is based on observations of an "old world" Temnothorax species collected in the United Kingdom, *Temnothorax albipennis*. Pratt (2005a) investigated how a "new world" species collected in the United States, *Temnothorax curvispinosis*, performed emigrations under similar conditions. He found that the behavioral algorithm followed by these two species is very similar, with the same stages of searching, assessing, canvassing, and accepting; similar division of labour between workers; and the use of a quorum threshold to mark the switch between tandem running and transport. What differed between the two species were the parameter values that determine the rates at which individuals switch between behavioral states. The most striking difference was in the quorum rule. *T. curvispinosis* required a higher threshold nest population before it switched from tandem running to transportation. Furthermore, on their first recruitment from the old nest to the new, individual *T. curvispinosis* have a larger quorum threshold than on later recruitments. As a result, more tandem runs were seen in the emigrations of *T. curvispinosis* than those of *T. albipennis*.

Honeybee House-hunting

Similarities in the behavioral algorithm employed during emigration are not limited to species of the same genus. As discussed in chapter 4, the movement decisions of cockroaches, fish, spiders, and other animals involve quorum-like responses to others (Sumpter & Pratt 2008). Honeybees are also thought to exhibit a form of quorum response during their emigration (Seeley & Visscher 2004b). Like *Temnothorax* ants, this quorum response is just one aspect of the algorithm the bees follow in choosing a new nest.

Seeley and Visscher have together with various colleagues studied the stages of the emigration of honeybee swarms. The emigration is initiated with the decision of a swarm of bees, including the queen, to leave the hive. This departure is remarkable in that it is preceded by relative calm, as the colony continues to function as usual. Then, during a period of about 10 minutes before a swarm of thousands or tens of thousands of bees departs the nest, there is a sudden and rapid increase in "piping," where an excited bee presses its thorax against a resting bee to produce a vibration in the passive bee's flight muscles, and "buzz-runs," where an excited bee runs around, pushing against other bees (Rangel & Seeley 2008). The bees lift off together and land in a nearby tree to form a bivouac-shaped swarm (Cully & Seeley 2004; Seeley 1995; Winston 1987).

From this swarm the bees collectively decide where to move. The process starts with scout bees searching for potential nest sites. Only a small proportion of the swarm members, probably around 5%, act as scouts or participate in the decision-making process (Seeley & Buhrman 1999; Seeley et al. 1979). When one of these scouts finds a nest site, she assesses its quality by walking and flying around the nest's interior (Seeley 1977). In contrast to *Temnothorax*, the time a scout takes to assess a nest is independent of the quality of the nest site and does not appear to involve the reconnaissance of other nearby sites (Seeley & Buhrman 1999; Seeley &

Visscher 2008). Instead, the scout will return to the swarm and perform a waggle dance to indicate the site's location to its nest-mates (Camazine et al. 1999; Seeley & Buhrman 1999). After the dance the scout will fly back and forth between the new nest site and the swarm, performing waggle dances at the swarm.

The swarm remains in its bivouac form for around two or three days before flying to a new nest site. During this time dances occur for a large number of potential sites, but directly before liftoff most, although usually not all, of the dances are for one of the available sites and it is to this site that the swarm flies (Camazine et al. 1999; Seeley & Buhrman 1999). Seeley & Buhrman (2001) offered a colony a choice between four mediocre and one superior nest. In four out of five trials the bees moved into the superior nest. The scouts are thus usually able to reach consensus for the best available site. However, in a small number of cases substantial dancing for other sites is seen directly before swarm liftoff (Lindauer 1955; Seeley & Visscher 2003). Such swarms have been seen to take off, but then hesitate as bees try to fly in different directions before finally returning to the original resting place of the swarm bivouac. Seeley & Visscher (2003) observed that after one such failed takeoff the bees regrouped and half an hour later lifted off again and flew in unison to one of the two dance-advertized sites.

Setting these occasional failures aside, the question is how the bees manage to reach consensus in the majority of cases. Visscher (2007) discusses four possible mechanisms that could promote consensus: individual comparison, dropout, competition, and inhibition. Visscher & Camazine (1999) performed an experiment in which scout bees that visited both of two available nest boxes were captured and removed from the decision-making process. The removal had no effect on the time it took the bees to reach consensus and move to one of the available sites, suggesting that individual comparison plays a relatively minor role in consensus building.

The most important mechanisms in reaching consensus appear to be a combination of dropout and competition. Although most potential sites elicit a dance response from the scout that first locates them (Seeley & Visscher 2008), over repeated visits dance intensity is higher for better quality sites (Seeley & Buhrman 2001). Furthermore, the intensity of dances for a site fades over repeated journeys between the nest and the swarm (Seeley 2003). The less rapid dropout for higher quality sites, coupled with competition for the limited number of available scouts that can follow dances, leads to a process of competitive exclusion (Britton et al. 2002; Myerscough 2003). Mathematical models of this process predict that the site at which the bees give up dancing for most slowly is eventually the focus of all dancing.

To decide whether there is sufficient support for the site for which they are dancing the scouts appear to use a quorum rule, similar to that employed by the ants (Seeley & Visscher 2003, 2004b). Once a particular potential nest site contains around 20 or so bees then waggle dancing is replaced by the same piping behavior seen prior to the swarm's initial departure from the old nest site. Piping is produced exclusively by scouts, and acts to warm up the other bees in the swarm in preparation for liftoff (Seeley & Tautz 2001; Visscher & Seeley 2007). As in the initial liftoff from old nest to bivouac, piping is accompanied by buzz runs, which also serve to activate bees resting within the swarm (Rittschof & Seeley 2008). Once activated, the swarm takes off and the scouts lead the entire swarm to the new nest site (see chapter 5 for details of how the small number of scout bees is able to lead the swarm).

Several models have converted the above description into a behavioral algorithm. Britton et al. (2002) and Myerscough (2003) both looked specifically at the dance dropout stage, showing how this dropout contributed to consensus. Janson et al. (2007) developed an individual-based model of the process of searching, dancing, and traveling back and forth between nest sites, investigating the role of scouting and how the colonies could cope with assessing nests at differing distances from the swarm bivouac. As well as dancing and dropout, Passino & Seeley (2006) include and investigate the quorum rule for the commencement of piping. As such, this is the most comprehensive individual-based model of honeybee house hunting.

Algorithm Analysis and Robustness

Once a behavioral algorithm is developed, the role of its various components can be tested. The algorithm can be analyzed and compared across systems. Indeed, rather than simply simulating algorithms in order to reproduce experiments, the algorithms can be studied to find out the principles that underlie them (Fewell 2003; Sumpter 2006).

An important way of gaining understanding of algorithms is between species comparison. Honeybees and *Temnothorax* ants both exhibit four main stages of the decision-making process: search (look for new nest sites both independently and through following dances/tandem runs), assess (evaluate discovered sites), canvass (dance/tandem run for an accepted site), and commit (bees pipe and perform buzz runs before leading the swarm to the new site, while ants transport to the new site). In both cases the switch from canvass to commit occurs when a quorum is reached at the perspective nest site. Pratt et al. (2002) showed that in *Temnothorax* ants quorum leads to a reduction in incidence of colony splitting. Similarly, Passino & Seeley (2006) used their model to demonstrate that the quorum improved decision-making accuracy. Passino et al. (2008) also note strong similarities between the mechanisms through which the bees reach a decision about which nest to move to and how populations of neurons reach decisions between different options.

Honeybees and ants differ in their respective requirements for speed and accuracy in decision-making. If a honeybee swarm takes off before a high level of consensus is reached then the swarm may split, an outcome that can prove fatal to those bees that do not move into the new nest site (Lindauer 1955). Splitting during emigration occurs in ants too, but colonies are later able to re-coalesce in the best of the available nests. Thus accuracy is more important than speed in the decision-making of honeybees. These differences may be reflected in differences in the way that quality is encoded in the recruitment by the two species. For the ants the assessment period is longer for lower quality nests, but once an ant has accepted a site her rate of recruitment is independent of site quality. Commitment by different ants to different nests can result in the quorum being reached for more than one site, and a higher degree of colony splitting.

For the honeybees, the assessment period is independent of quality but dancing is more vigorous for better quality sites. This quality-based recruitment, combined with more rapid dropout of dancers for inferior sites allows the scouts to reach near consensus before liftoff (Myerscough, 2003). Thus, while dropout might lead to slower decision times, it provides an improvement in decision accuracy.

How behavioral algorithms are employed in different situations can give further insight into the trade-off between speed and accuracy. Pratt & Sumpter (2006) looked at how ants tune their behavioral algorithm to different challenges. As well as moving nest when their current nest is destroyed, *Temnothorax* ants are known to move up the "housing ladder" as better nest sites are made available to them (Dornhaus et al. 2004). Pratt & Sumpter compared how the ants migrated under these "unforced" conditions, when the ants live in a poor quality nest and one or more better quality nests becomes available, with "forced" conditions, where their current nest is destroyed. When choosing between a good and a mediocre nest, colonies showed different behavior depending on the urgency of their need to move. In the unforced situation colonies took a long time to emigrate, but they more often chose the better of the two available nests. In forced emigrations, colonies moved much faster but often made poor choices, splitting their population between the good and mediocre nests or even moving entirely into the inferior one.

Pratt & Sumpter showed that while the speed and accuracy of decision-making was tuned to the circumstances of forced and unforced emigrations, the behavioral algorithm employed by the ants was the same in both cases. The four stages of searching, assessing, canvassing, and finally accepting and transporting after a quorum is met were seen in both forced and unforced emigrations. The difference in speed and accuracy in the two different circumstances resulted from an increased urgency on the part of individual ants in forced emigrations. The rates of leaving the old nest to search and of accepting a newly found nest were larger and the quorum threshold was lower in forced emigrations. The ants changed the parameters of the algorithm, but the algorithm itself remained unchanged. The ants appear to have evolved a single algorithm, which can be tuned to differing requirements of speed and accuracy. Franks et al. (2003a) further showed that the ants may well employ the same algorithm in even more desperate situations. By adding formic acid to the colony they induced an emergency move with an even lower quorum threshold than a normal forced emigration.

The behavioral algorithm adopted by the ants appears remarkably robust to differences in the number of and distance to the available nests. Pratt (2008) compared emigrations to nearby and distant nests. He found that emigrations to distant nests involve more tandem runs, with the followers of these tandem runs responsible for more transportation, than in emigrations to nearby nests. In terms of the behavioral algorithm, the quorum is reached slower for distant nests, leading to a later switch to transportation and a greater effort in tandem runs that inform other ants where the new nest site is. The quorum rule thus tunes the amount of tandem running to the level of difficulty in finding a new nest. Model simulations confirm this interpretation of the data (Sumpter and Pratt, unpublished results).

Franks et al. (2008) examined how the ants cope when offered one nearby nest of poor quality and a distant nest of better quality. Even when the better nest was nine times farther away than a poor quality nest the colony successfully moved into the better nest. Individual comparison and the assessment delay, which allowed ants that had found the nearby poor quality nest to find the far away good quality nest before they commenced transportation, played important roles in decision-making. In cases where transportation did commence to the poor quality nest, it was quickly superseded by recruitment to the better nest. Similar results have been found when a better quality nest is introduced once an emigration has already commenced to a mediocre nest. The ants are often able to swap mid-emigration to the better nest (Franks et al. 2007).

While the algorithm is robust to different environmental conditions, how robust is it to changes in parameter values? One interpretation of the similarities in the algorithm but differences in parameter values between *T. albipennis* and *T. curvispinosis* is that the algorithm offers a degree of robustness that is independent of particular parameter values

(Pratt 2005a). Natural selection acts to shape the algorithm, but parameter values are not strongly constrained. A similar argument could also explain why no consistent relationship has been found between the size of the quorum and the size of emigrating colonies (Dornhaus & Franks 2006; Franks et al. 2006; Pratt 2005a). Such robustness has also been hypothesized as a design property of the gene networks that regulate development (von Dassow et al. 2000).

Formalizing Individual-based Models

Often the presentation of individual-based models in the scientific literature consists of a flow chart or written description without a precise mathematical specification of the model. The lack of mathematical specification can be contrasted with differential equation or simpler stochastic models where papers clearly specify the equations underlying the model, allowing other researchers to reproduce the results.

One of the ambitions of building the individual-based model of *Temnothorax* migration was to provide a reliable tool against which to predict and understand future experimental results. To fulfill this aim Pratt et al. (2005) provided an unambiguous specification of the model in terms of a specification language WSCCS. Box 9.B introduces some of the basic aspects of WSCCS. This specification language was proposed by Tofts for modeling various aspects of social insect organization (Tofts 1991, 1993, 1994).

There is a strong advantage to using a specification language if all researchers adopt the same language. Unfortunately, there are also several disadvantages to using formal specification languages in model building. First, keeping a clear specification can become burdensome when attempting to develop in parallel a number of different models of the same system. For example, previous to the Pratt et al. (2005) individual-based model of Temnothorax emigration, Pratt et al. (2002) had written a differential equation model of the quorum mechanism. This latter model proved extremely powerful in understanding why the ants employed a quorum threshold. While it was possible to see the quorum model as a simplification of the full individual-based model under a certain set of assumptions, the formal steps required to make this simplification were cumbersome and revealed nothing new.

A second disadvantage of formal specification is that particular specification languages are not always designed to address all modeling problems. WSCCS is not good for representing processes that occur on very different time scales, or problems that are spatially explicit. For *Temnothorax* emigration, several researchers have developed their models in

Box 9.B Process Algebra Models

Process algebras were first proposed to aid the analysis of the performance of distributed computer systems (Bruns 1997; Milner 1989). They allow formal specification of the individual components, which make up a system, and then provide means for proving properties of component interactions. Standard process algebras are designed to examine conditions under which a system will fail or attempt to prove that a system will never fail. As such they look at properties of a system that are time-independent and are unaffected by stochastic variations. In modeling insect societies it is usually the timing and probability of events that are of greatest interest to us. To this end, Tofts (1991, 1993) developed a probabilistic and time-dependent version of one of the most widely-used process algebras, Calculus of Communicating Systems (CCS). Toft's Weighted Synchronous Calculus of Communicating Systems (WSCCS) allowed him to specify models where ants were the components, and then answer questions about properties of the colony they composed.

A simple example is as follows. We can define an ant waiting at a nest as follows:

$$\text{ATNEST} \quad = s: \sqrt{.}\text{SEARCH} + (1 - s): \sqrt{.}\text{LOOKCALL}$$

$$\text{LOOKCALL} \quad = \omega:call.\text{FOLLOW} + 1: \sqrt{.}\text{ATNEST}$$

This definition can be interpreted as follows. An agent in state AT-NEST will with probability s enter the state SEARCH and with probability $(1 - s)$ enter the state LOOKCALL. The symbol $\sqrt{}$ denotes that one time step of the simulation will pass when the state is updated. The agent LOOKCALL also has two possible actions, *call* or $\sqrt{}$. ω denotes that the action *call* is prioritized over $\sqrt{}$. These actions can only be understood in the context of the interaction of two or more agents. In particular, in order for the action *call* to be performed it is required that there is another agent performing the complementary action <u>*call*</u>. For example, if we define two further agents:

$$\text{GIVECALL} = \omega:\underline{call}.\text{LEAD} + 1: \sqrt{.}\text{GIVECALL}$$

$$\text{WALK} = 1: \sqrt{.}\text{WALK}$$

(Box 9.B continued on next page)

when these three agents are defined in parallel, written as:

$$\text{LOOKCALL} \times \text{GIVECALL} \times \text{WALK}$$

then on one time step these agents will become

$$\text{FOLLOW} \times \text{LEAD} \times \text{WALK}.$$

On the other hand, the parallel definition

$$\text{LOOKCALL} \times \text{WALK} \times \text{WALK}$$

will become

$$\text{ATNEST} \times \text{WALK} \times \text{WALK}.$$

Thus if two parallel agents wish to perform the complementary actions, *call* and <u>*call*</u>, then this takes priority.

Since the idea of WSCSS is to give a formal definition of agents, the above informal discussion does not give an unambiguous explanation of how process algebras are defined. A more complete description can be found in Tofts (1991, 1993) and Sumpter et al. (2001). WSCCS can, however, be used to provide formal definitions of complicated state-based models of animal interactions. It can also be used to develop Markov chain and differential equation representations of these models (Sumpter et al. 2001). In the supplementary material of Pratt et al. (2005) a full description is given of the *Temnothorax* nest choice model in WSCCS.

order to investigate different aspects of the emigration. These range from differential equation to various individual-based models. Thus, although the ideal would be for everyone to use a consistent language for specifying their models, in practice this is a very difficult aim to fulfill. The burden of making every model consistent outweighs the advantage of rigorous comparison between models.

The lack of agreed-upon framework for developing individual-based models and the difficulty in measuring the large number of parameter values discussed at the end of the Social Insect Foraging section in this chapter, are the two major reasons for a limited acceptance of this type of modeling approach (Grimm & Railsback 2005). Individual-based models are often viewed as ill-defined and unreliable by researchers versed in the use of differential equations or other mathematical approaches. This is unfortunate, because in some sense they are the only tool available in the study of truly complicated systems. Exactly how the lack of standards and reproducibility in individual-based modeling can be overcome in the future is unclear, but it remains an important problem.

In this chapter I have focused on individual- and state-based models in the context of social insect foraging and migration. These examples should serve primarily as case studies of the successes and limitations of this type of approach. Indeed, individual-based models are by no means limited to these applications. Agent-based modeling is a powerful tool for understanding the social sciences (Edmonds et al. 2008; Miller & Page 2007), ecology (Grimm & Railsback 2005) and microbiology (Ferrer et al. 2008).

Dimension Reduction

If real world systems consist of large numbers of variables and parameters, and models with large number of variables and parameters are unwieldy, then how can we hope to model complicated systems? One answer to this question lies in the very art of modeling: to find a way of expressing the key elements of a system in only a few well-defined variables. It is the art of modeling to try to produce a model that is as "simple as possible but no simpler," as the quote attributed to Einstein goes.

The number of variables in a model is often referred to as the model's dimension. With the exception of the models discussed in this chapter, most of the models in this book have a small dimension, e.g., a small number of variables describing the number of ants visiting a particular feeder or the proportion of individuals that scrounge or produce food opportunities. In cases where the model's dimension is higher, for example, when we have a pair of variables for the position and speed of each particle in the SPP models of chapter 5, we try to derive a smaller number of variables, such as the instantaneous alignment or average neighbor distance, that somehow characterize the group. The aim here is to characterize large- or infinite-dimensional models by a much smaller number of variables. If we can characterize a system by a model that has only a small number of dimensions, then it becomes easier to make mathematical predictions about its behavior.

Complications should never be overestimated. When viewed at certain spatial and temporal scales very complex individuals can produce very simple group level dynamics. For example, though a highly complex algorithm may have brought shoppers to town in the first place, the number of people passing by a point on a quiet shopping street during a five-minute interval is likely to be randomly distributed. This prediction is based on the "law of small numbers," that independent low frequency events in a large population follow a Poisson distribution (Bortkiewicz 1898). While on very small time scales there are socially enforced gaps between people and on very large time scales there are patterns determined by shops opening and closing, on the time scale of an hour on a Monday morning pedestrians pass by more or less at random. Once a large number of factors begin to influence behavior, the complex begins to seem simple again.

Another example of a single distribution characterizing large numbers of independent individuals is the central limit theorem. In box 4.B in chapter 4, I show how the normal distribution, which is defined by its mean and variance, characterizes the sum of the actions of n independent individuals. The "law of small numbers" and the central limit theorem are just two examples of mathematical results that allow a system of high dimension to be simplified to one of low dimension. The mean-field approximations used to derive differential equation models in box 9.A are another. Mathematical techniques such as moment closure (Keeling & Ross 2008), equation free methods (Kevrekidis et al. 2004) and others all act to reduce the dimension of complex models and bring clearer analytical understanding of systems (Sethna 2006; Sornette 2004).

Complications should not be underestimated either. The fact that I have written seven chapters on models with a small number of dimensions and one chapter on those with high dimension should not suggest that most problems in collective animal behavior can be modeled using a small number of variables. The decision to study a particular scientific problem is based not only on its intrinsic importance, but also upon whether we believe we can make progress solving it. Low-dimensional models are mathematically tractable. If such a model can be applied in understanding a system, it becomes more likely to be the subject of experimental research. While a large number of mathematical techniques have been discovered to reduce biological systems to lower dimensions, there is no a priori reason that the majority of these systems should be low dimensional. Indeed, while we can accept the idea that the complex is sometimes simple, our everyday experience tells us that the biological world is truly complicated.

— Chapter 10 —

The Evolution of Co-operation

A fundamental question about all forms of collective animal behavior is how they evolved through natural selection. At various points in this book I have turned to arguments based on individuals adopting or evolving behaviors that increase their own fitness to explain or make predictions about group behavior. For example, group size distribution was described in terms of individuals attempting to join a group of a size that maximizes their fitness (chapter 2); foraging birds were described as balancing searching for food themselves with copying others (chapter 3); consensus decision-making and synchronization were described in terms of individuals co-ordinating so they can benefit from acting together (chapters 4 and 7). While such functional arguments are not the only way to understand the behavior of groups (and indeed have played a secondary role to mechanistic explanations in the other chapters of this book), they are an essential part of biology. This chapter gives an overview of how functional reasoning can be applied to collective animal behavior.

The theory of natural selection is grounded in the idea that those individuals exhibiting a behavior that provides them with higher than average fitness pass their genes, and thus their particular behavior, on to future generations. It is this idea that provides the basic assumption of evolutionary game theory models: those individuals adopting a strategy that provides them with higher than average fitness will increase in the population, while those with lower than average fitness will decrease. Despite the simplicity of this underlying assumption, these models have proved extremely powerful in predicting when co-operation between animals will evolve (Dugatkin & Reeve 1998; Maynard Smith 1982). As a result of this success, a vast literature has arisen on the evolution of co-operation, both theoretical and experimental. The size of this literature makes it difficult to give a concise account of how different models and experiments relate to one another. There is, however, an increasing consensus of how co-operation should be discussed in evolutionary biology (Clutton-Brock 2002; Foster et al. 2006; Lehmann & Keller 2006a, 2006b; West et al. 2007). In this chapter, I follow this consensus, and categorize how collective and co-operative behavior can evolve between non-relatives in four different ways: through parasitism, mutualism, synergism, and repeated interactions. In doing so I discuss evolutionary game theory, which has become a central modeling tool in understanding co-operation between non-relatives.

A complication arises when we consider the evolution of altruism. Altruism is defined as individuals paying a cost greater than any resulting benefit to their average lifetime reproductive success while providing a benefit to the lifetime reproductive success of others. Under natural selection, an individual adopting an altruistic strategy will suffer a drop in fitness and will be less likely to directly contribute offspring to the next generation. By introducing genetic relatedness into evolutionary game theory models, however, we see that the inclusive fitness equation or Hamilton's rule provides a good predictor of when altruism can evolve. Altruism can evolve when individuals help their relatives, thus indirectly passing their genes to future generations. This chapter should thus provide a broad classification of many of the collective behaviors discussed in this book as arising from a combination of four distinct forms of co-operation (parasitism, mutualism, synergy, and repeated interactions) and altruism arising from inclusive fitness.

Evolutionary Game Theory

Evolutionary game theory models describe how selection acts on behavioral strategies. Over repeated generations a strategy will increase in the population if it receives a higher than average payoff or fecundity, but decrease if it receives a lower than average payoff or fecundity. The biological interpretation of these arguments is different when considering evolution through natural selection or individual decision-making (Dugatkin & Reeve 1998). Under Darwinian natural selection, we assume that a genotype encodes for a particular strategy throughout an individual's lifetime and ask whether this genotype will increase or decrease in the population over generations. Those genotypes that provide lower fecundity die without reproducing and those providing higher fecundity produce offspring that fill their places. Under individual decision-making, we consider an individual that can change its strategy during its lifetime in response to its mistakes. This individual plays its strategy. If it then gets a payoff higher than average it keeps the strategy; otherwise, if its payoff is lower than average, it changes strategy to one that will improve its payoff.

Evolutionary game theory assumes the following lifecycle for individuals. An infinitely large population of individuals undergo the following stages:

1. Individuals form groups of size N. Individuals are distributed entirely at random between groups, i.e., dispersal is global.
2. Each individual has a behavioral strategy, s_i. The payoff or fecundity of each individual i is determined by its own strategy and that of all other individuals in the group, i.e., $f(s_i, s_1 \ldots s_N)$. No reproduction occurs while individuals are within the group.
3. Individuals then leave the group and a law of selection is applied to them: each strategy's contribution to the next generation is proportional to its fecundity relative to the average fecundity of the entire population (i.e., not just those in the group of size N). This contribution to the next generation is also known as the individual's fitness.

In many evolutionary game theory models it is often further assumed that $N = 2$ (Maynard Smith 1982). For $N > 2$ these assumptions are the same as "group selection" models (Nunney 1985; Wilson 1983). Provided $N < \infty$, the probability of repeatedly interacting with the same individual on consecutive generations is zero, as is the interaction probability for two related individuals, e.g., individuals with the same parents.

Some of these assumptions can be relaxed and the results of these models remain the same. For example, Nunney (1985) shows that a population interacting with local neighbors in a continuous space, rather than discrete groups, but dispersing globally before reproduction gives similar predictions as those from the lifecycle above. What cannot be relaxed, however, is the assumption that there is a zero probability of interacting with the same individual twice or of interacting with relatives. These two cases are dealt with separately below in the sections on repeated interactions and inclusive fitness, respectively. For a good discussion and justification of the other assumptions underlying these models see Grafen (1984), Nunney (1985), and Dugatkin & Reeve (1994).

Evolutionary games where $N = 2$ and individuals choose between two distinct strategies can be expressed in terms of a payoff table. Table 10.1 gives such a payoff table for interactions between "co-operators" and

TABLE 10.1
Payoff table for two player evolutionary game

Focal/Partner	Co-operate	Defect
Co-operate	$B - C + E$	$D - C$
Defect	B	0

The values in the table determine the fitness gained by the focal individual as a function of its own strategy and that of its partner.

225

Box 10.A Two-player Discrete Strategy Evolutionary Games

In addition to the assumptions in section 10.1, I further assume that $N = 2$ and there are only two discrete behavioral strategies, called co-operate and defect. Interactions in these pairs give a pay-off, i.e., fecundity or number of offspring produced, which can be expressed in a two-by-two payoff table (table 10.1). A focal individual adopting the co-operative strategy always pays a cost C and always confers a benefit B to its partner. Depending on the strategy of the partner the focal individual receives a *direct benefit D*, if the partner defects, or an *extra benefit E*, if the partner co-operates. A focal defector pays no cost but receives the benefit B if its partner is a co-operator. Note that B, C, D, and E are assumed to be positive constants that do not depend upon the frequency of co-operators or defectors in the population as a whole.

The payoff to an individual depends on the strategy of their partner. Since pairs are selected entirely at random from an infinite population, the probability that a focal individual interacts with a co-operative individual is equal to the proportion of co-operators in the population as a whole, denoted x. Thus the expected payoff of a co-operator is

$$x(B + E - C) - (1 - x)(D - C),$$

and the expected payoff of a defector is xB.

These are the frequency-dependent payoff functions, which determine how many offspring individuals produce. The average payoff of the population is

$$x(x(B + E - C) - (1 - x)(D - C)) - (1 - x)xB.$$

Natural selection implies that the proportion of individuals with higher than average payoff will increase and the proportion of individuals with lower than average payoff will decrease in the population. Thus the fitness of an individual can be defined as being equal to its payoff divided by the average payoff of population, i.e.

$$f(x) = \frac{x(B + E - C) + (1 - x)(D - C)}{x(x(B + E - C) + (1 - x)(D - C)) + (1 - x)xB}.$$

We can then write the following expression for the rate of change of the proportion of co-operators in the population:

$$\frac{dx}{dt} = x(\text{payoff of a co-operator} - \text{average payoff of population}),$$

which in this case is equal to

$$\frac{dx}{dt} = g(x) = x(1-x)\big(x(E-C) + (1-x)(D-C)\big).$$

Solving this equation as $t \to \infty$, i.e., $dx/dt = 0$, gives us the conditions under which co-operation is selected for.

Solving $dx/dt = 0$ gives steady states at $x_* = 0$, $x_* = 1$, and $x_* = (C-D)/(E-D)$. These correspond respectively to a population of all defectors, all co-operators, and a mixture between co-operators and defectors. By differentiating $g(x)$ with respect to x and evaluating at x_* we find the conditions under which the addition of a small number of co-operators or defectors will lead to the growth of that strategy in the population away from the steady state. These steady states are said to be evolutionarily stable to such small perturbations whenever $g'(x_*) < 0$. For example, $g'(0) = D - C$ so all defect is stable if $C > D$. Similarly, all co-operate is stable if $E > C$. The mixed strategy steady state, $x_* = (C-D)/(E-D)$, exists (i.e., lies between 0 and 1) and is evolutionarily stable if $D > C > E$. The mixed strategy steady state exists but is unstable if $E > C > D$. These results are summarized and categorized in figure 10.1.

"defectors." In this game we assume that a focal co-operator pays a cost C and confers a benefit B to its partner. If the partner defects the focal co-operator still receives a *direct benefit D*, while if the partner co-operates it receives the benefit B plus an *extra benefit E*. A focal defector pays no cost but receives the benefit B if its partner is a co-operator. The model resulting from the above assumptions and this payoff table is analyzed in box 10.A.

The distinction between the different types of benefits B, D, and E lies at the heart of understanding the evolution of co-operation. Figure 10.1 shows how the relationship between the parameters D and E and the cost C determines the evolutionarily stable proportion of co-operators and defectors in the population. The figure gives four qualitatively distinct evolutionary outcomes, which I call parasitism, mutualism, synergy, and "failed" altruism. I now examine each of these scenarios separately, also

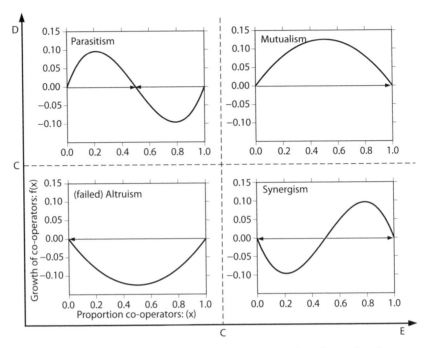

Figure 10.1. Outcome of two-player evolutionary game. Shows how the cost-benefit parameters, C, D, and E determine the evolutionarily stable strategies in two-player, two-strategy games. The axes of the main figure are the benefits E and D. Each of the four panels within the figure shows one of four qualitatively different forms of the fitness function $g(x)$, given in box 10.A. The arrows show how the amount of co-operation in a population will change, given a particular proportion of co-operators in the population. If $D > C$ but $E < C$ then the outcome is parasitism, a single evolutionarily stable state where the population consists of a mixture of co-operators and defectors; if $D > C$ and $E > C$ then the outcome is mutualism, the evolutionarily stable state is to all co-operate; if $D < C$ and $E > C$ then the outcome is synergism, where there are two evolutionarily stable states one corresponding to all defect and one to all co-operate; finally, if $D < C$ and $E < C$ then the evolutionarily stable state is to all defect. This last case is classified as failed altruism, because were individuals to co-operate in such a situation their actions would be altruistic and result in negative direct fitness for an individual.

discussing how the results for these two-player, two-strategy games relate to N-player and continuous strategy games. It turns out that two-player games provide quite a general classification of the possible forms of co-operation, even when we add these further complications.

Parasitism

It's Sunday morning, the living room is a mess, but the newspaper has just arrived. You and your partner have the option of either tidying the living room or sitting down and enjoying the lifestyle supplement. The decision can be interpreted in terms of the costs and benefits in table 10.1. Assume that D is the direct benefit you gain from having a tidy living room, but you pay cost C in the time you spend tidying. Further assume that the benefit of having a tidy room is greater than the cost of cleaning, so $D > C$. On the other hand, if your partner tidies up then you can defect and get the benefit B of reading and having a tidy room. If you tidy up together then you save some time E, but because tidying up always takes some positive amount of time, $C > E$.

Identifying $D > C > E$ in figure 10.1 shows that a unique evolutionarily stable state exists for individuals that wake up in a different house with a different partner every Sunday. A single defector in a population of co-operators benefits from the work of the others, leading to an increase in defection. However, when everyone defects then they all receive the lowest possible payoff and it pays for an individual to co-operate instead. The proportion of individuals who co-operate evolves to $(C - D)/(E - D)$. At this point no one individual can do better on average by changing strategy. The lack of stability of both pure co-operation and pure defection leads to a "compromise" of some individuals that work and some that "free ride" or parasitize the work of others.

Social Parasitism in Groups

Social parasitism is of relevance to a large number of social and biological situations. The producer-scrounger game discussed in chapter 3 in box 3.B describes a situation where a focal foraging bird can either search for food itself (produce) or watch others search and share their finds (scrounge). The strategy an individual chooses depends on the strategies of others: it pays to scrounge in a population of pure producers and it pays to produce in a population of pure scroungers. The evolutionarily stable strategy, of a mixture between some scroungers and some producers, is widely observed in the behavior of foraging birds (Producers and Scoungers, chapter 3).

A term that often arises in describing N-player social parasitism is the "Tragedy of the Commons" (Hardin 1968). Imagine a commons pasture on which any individual can place their cattle. This pasture has a limited capacity and having cattle above this capacity causes the pasture to deteriorate in quality. A tragedy arises because the benefit of adding a cow goes directly to the owner, while the cost in deterioration is shared among everyone. Thus even when the pasture is over capacity, it can pay for an individual to add a cow. Ultimately, the pasture will yield an average payoff below that yielded at capacity. The same is true in the producer-scrounger game: at the evolutionarily stable state the average intake of each forager is lower than it would be if they all searched independently. And again for vigilance: flocks are less vigilant than would optimize their energy intake (Fernandez-Juricic et al. 2004b). And again in the group size paradox in chapter 2: it pays solitary individuals to join a group even if this group size will become suboptimal for those already in it.

Hardin uses the "Tragedy of the Commons" to argue for regulation of individual freedom, since individual freedom will lead to disadvantage for all. While this argument has some validity, the outcome of social parasitism is not as bleak as it might at first seem. Provided there remain direct benefits for co-operating, i.e., $D > C$, then the tragedy does not lead to everyone receiving the lowest possible payoff of zero. When no one else is producing or being vigilant, it always pays to co-operate. The fact that in a fully co-operative society it pays for some individuals to defect does not imply that everyone will defect. We should not be surprised when we see animals or humans co-operating despite parasitism of their efforts by others, but should first consider what direct benefits are gained by the co-operators (Clutton-Brock 2002; Griffin & West 2002).

Continuous Strategies

In the model in box 10.A we assume that there are two distinct strategies, one corresponding to co-operate and the other to defect and study how the proportion of the population adopting each strategy evolves. Another approach is to have a continuous strategy, expressing the level of investment made in co-operation. For example, we can think of each individual in our population of living room occupants as investing a level, p, in tidying up.

It is tempting to think of the proportion of the population and the level of investment as being the same thing, and conclude that individuals with continuous strategies will adopt a level of co-operation equal to the evolutionarily stable strategy in the discrete strategy game. The problem with this interpretation is that a population where all individuals have strategy p_* can be, although not always is, invaded by two strategies that lie on either side of p_* (Vincent & Brown 1984). Ultimately, a sequence of invasions may lead to the population again dividing into having two distinct strategies, one that always co-operates and one that always defects.

The above observation is important because animals are often faced with decisions not simply whether to co-operate or not, but how much they want to invest in a co-operative behavior, such as guarding or searching for commonly exploitable food. The importance of continuous strategies has been widely acknowledged since evolutionary game theory was first developed (Zeeman 1981). Geritz et al. (1998) provided an elegant framework for determining when evolution will converge on a particular strategy. Box 10.B shows how this framework, often referred to as adaptive dynamics, can be used to classify an evolutionary game with continuous strategies in which a focal individual who invests p in a co-operative behavior receives payoff

$$P(p,q) = B(p,q) - C(p),$$

where q is the level of investment by its partner (Doebeli et al. 2004). Figure 10.2 shows that the population will first move toward adopting a single strategy. If this strategy is not pure defect or pure co-operate then, depending on the exact form of $B(p,q)$ and $C(p)$, the population will branch into two distinct types, one type that invests more (co-operators) and one that invests less (defectors) in co-operation than those at the mixed singular strategy (figure 2a). Alternatively, the population will remain at this mixed strategy where all individuals invest the same amount in co-operation (figure 10.2b). For continuous strategy games it is this final outcome that is the evolutionarily stable strategy, which cannot be invaded by mutant individuals adopting small changes to their strategy.

The main biological message of the continuous strategy game is that even when individuals have an opportunity to tune a level of investment in a co-operative activity, we can expect discrete strategies. The co-operators produce most of the benefit and pay most of the cost and the defectors parasitize their efforts (Doebeli & Hauert 2005; Doebeli et al. 2004). This could explain why some individuals seem to specialize in apparently costly behaviors such as searching for food (Barnard & Sibly 1981), while others apparently reap only the benefits of these behaviors. It is important to note, however, that the co-operators in such scenarios still derive a direct benefit from co-operation. By definition, the co-operating individuals cannot increase their payoff by defecting and are thus acting in their own selfish interests.

Mutualism

If there are positive benefits to performing a co-operative behavior independent of the actions of a partner, then co-operation can evolve despite the existence of associated costs. If in table 10.1 there is an extra benefit that cancels out the cost, i.e., $E = C$, the payoff for both co-operating is $B + E - C = B$. Now there is no longer a positive incentive to defect. If E continues to increase, then the total benefit becomes more than the sum of the separate actions and there are now mutual benefits to both co-operating, i.e., $B + E - C > B$. The extra benefit of working together

Box 10.B Two-player Continuous Strategy Games

Doebli et al. (2004) describe an evolutionary game theory model where $N = 2$ and players meet in pairs. Each player invests a level between 0 and 1 in co-operative behavior. They assume p is the level investment of the focal individual, q is the level of investment by its partner and the focal individual receives payoff

$$P(p,q) = B(p,q) - C(p).$$

Consider a population in which every individual adopts the same strategy q, apart from a rare mutant that adopts strategy p that is slightly different than q. Whether a mutant invades and the resident strategy changes through time is determined by the selection gradient

$$D(q) = \frac{\partial}{\partial p}\left(P(p,q) - P(q,q)\right)\Big|_{p=q}$$
$$= B'(q,q) - C'(q).$$

Steady states q_* of $D(q_*) = 0$ are said to be convergent stable when $D'(q_*) < 0$, where the derivative here is taken with respect to q. The convergent stable states are the points toward which the population first evolves in figure 10.2.

The fact that q_* is convergent stable does not imply that it is the final resting point of evolution. It is possible that two mutants, one with a slightly larger investment than q_* and one with a slightly smaller investment than q_* can simultaneously invade the resident q_*. Doebli et al. (2004) show that the convergent stable state q_* is the final stopping point of evolution if $B''(q_*,q_*) - C''(q_*) < 0$, where the partial derivative here is taken twice with respect to p and then evaluated at $p = q = q_*$. In this case, every individual makes the same investment q_* in co-operation (e.g., figure 10.2b). If $B''(q_*,q_*) - C''(q_*) > 0$ then q_* is said to be an evolutionary branching point and the population separates into individuals that adopt more co-operative and less co-operative strategies (e.g., figure 10.2a). Details of the above classification of steady states is given by Geritz et al. (1998).

Doebli et al. (2004) looked at non-linear benefit and cost functions of the form

$$B(p,q) = b_2(p + q)^2 + b_1(p + q)$$

$$C(p) = c_2 p^2 + c_1 p$$

and found a steady state at $q_* = (c_1 - b_1)/(4b_2 - 2c_2)$, which exists and is convergent stable when $4b_2 - 2c_2 < c_1 - b_1 < 0$, and which exists and is evolutionarily stable when $c_2 > b_2$. By insisting that investment in co-operation must be between 0 and 1, two further steady states are created at these points.

There are five different evolutionary outcomes of this model, examples of which are shown in figure 10.2. Four of the five outcomes are identical to the outcomes of the discrete strategy model shown in figure 10.1—some individuals defect, some co-operate (figure 10.2a, parasitism); depending on initial population all defect or all co-operate (figure 10.2c, synergy); all defect (figure 10.2d, failed altruism); or all co-operate (figure 10.2e, mutualism). The additional outcome (figure 10.2b) arises when the mixed strategy q_* is evolutionarily stable. In this case individuals balance their own investment in co-operation so as not to be exploited. The continuous model thus distinguishes the cases where the population all adopts the same strategy or splits between two strategies, but in both these cases the interactions are social parasitic.

outweighs any cost. In figure 10.1, I label this scenario mutualism. Both E and D are greater than C and under all circumstances it pays to co-operate.

Mutualism provides an explanation of many different behaviors by animal groups: animals aggregate in the hope that another individual in a group will be eaten by a predator (chapter 2), pigeons flying together benefit from each other's directional information (chapter 5), and so on. In many ways, mutualisms provide a null hypothesis for co-operative behavior. If we see an animal performing a costly behavior that benefits another individual, the first question we can ask is what benefits it gains itself from the action. If it gains irrespective of the actions of its partner then the interaction is mutualistic.

Despite providing a somewhat obvious reason for individuals to co-operate, mutualisms are sometimes overlooked. This can be because the benefits are not immediately clear or the costs are overestimated. Clutton-Brock (2002) emphasizes that, when estimating these costs and benefits, the physiological state of the individual performing them must be taken into account. This state can differ between individuals. Thus

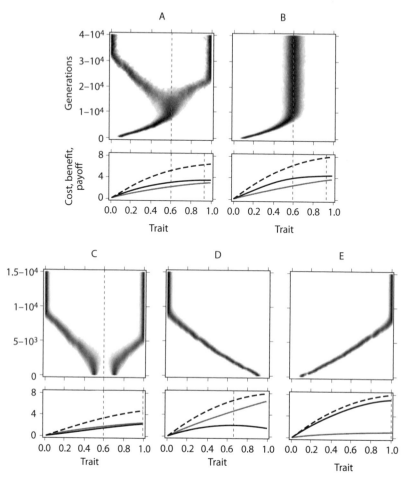

Figure 10.2. Outcomes of a simulation of a continuous investment two-player game (reproduced from M. Doebeli, C. Hauert, & T. Killingback, "The Evolutionary Origin of Cooperators and Defectors," October 2004, *Science* 306, 859–862, fig. 1 reprinted with permission from AAAS). Top row shows the evolutionary dynamics of the strategy distribution; darker shades indicate higher frequencies of a strategy. Singular strategies (dashed vertical lines) are indicated where appropriate. Bottom row shows the cost $C(x)$ (dotted line), benefit $B(2x)$ (dashed line), and the mean payoff $B(2x) - C(x)$ (solid line). The dash-dotted vertical line indicates maximal mean payoffs. (A) Social parasitism with evolutionary branching; (B) Social parasitism with an evolutionarily stable singular strategy; (C) Synergy with two evolutionarily stable strategies; (depending on the initial conditions, the population either evolves to full defection or to full cooperation, two distinct simulations shown); (D) Failed altruism; (E) Mutualism. See Doebli et al. (2004) for details and parameter values.

while co-operative actions may have substantial energetic costs, they may be performed by individuals with high energy reserves. For example, as discussed in chapter 6, meerkats spend more time guarding their collective nest when they are well fed. While guarding may be costly to a hungry meerkat, it is beneficial to one with a full stomach (Clutton-Brock et al. 2001, 1999). It thus becomes mutually beneficial for meerkats to guard their nest and take turns (Bednekoff 1997; Foster 2004).

In mutualisms, the balance of power is often tilted toward one individual (Beekman et al. 2003). For example, carpenter bee queens usually find nests on their own, but are sometimes usurped through a violent struggle by another queen (Hogendoorn & Velthuis 1999). The usurper then lays all subsequent eggs. Ironically, the stage is now set for mutualism between these one time opponents. The usurper can benefit if the original foundress stays to guard the nest, but must provide an incentive for the foundress to stay. She can do this by not destroying all of the eggs laid by the foundress. While this small concession may provide only a small benefit to the foundress, it is greater than the probable benefit arising from an attempt to establish a new nest (Dunn & Richards 2003). Carpenter bee nest founding provides just one example of a "transactional concession" offered by a dominant individual to one or more sub-dominants. These concessions make it mutually beneficial for the sub-dominants to co-operate (Reeve & Keller 2001). Unlike in table 10.1, the dominant and sub-dominants have different payoff tables and the dominant uses concessions to manipulate the payoff table of the sub-dominants. This manipulation ensures that both have a benefit in co-operating, independent of the future actions of the other. Thus, while power and mutualism might appear unlikely partners in animal conflicts, as in human affairs, they lie at the heart of many co-operative group behaviors.

It is particularly important when considering mutualisms to observe that the benefit B has absolutely no role in the evolutionary dynamics (box 10.A). This is somewhat counter-intuitive since it means that we can expect individuals to evolve to confer arbitrarily large benefits on others, giving a strong appearance of altruism. What should be borne in mind here is that for a particular act to benefit a partner, it usually incurs a cost to the actor. For example, giving your food to another individual might benefit them to degree B but will cost you to degree C. If the act of giving food provides no direct or extra benefit (i.e., $D = E = 0$), then however small the cost is of giving food and however large the benefit is to another, it should not evolve.

Despite the above limitation, we can expect to see acts where individuals confer huge benefits to others. In particular, acts, which incur costs that are slightly smaller than the direct benefits they provide, i.e., C is slightly smaller than D, may provide extremely large benefits to others. A human example of such a mutualism would be a confident swimmer jumping in a river to save a stranger's life. The confident swimmer is unlikely to drown (small C) and will reap a benefit in terms of reputation or simply from the vigorous exercise (D slightly bigger than C), but in doing so will provide an extremely large benefit B to the drowning stranger. However happy the drowning stranger is after the rescue, by the evolutionary game theory definition, this is no more altruistic than if the lifesaver had trained for, entered, and won a swimming contest.

Synergism

Much of this book is about how non-linear interactions between individuals can produce patterns that a single individual could not achieve alone. The chemical trails of rats and ants, cliff swallow food calling (chapter 3), bark beetles attacking trees and caterpillars building tents (chapter 7), all involve the mass co-operation of many individuals that ultimately improves individual efficiency. In this sense their co-operation could be described simply as mutualistic. However, there is an aspect of these co-operative behaviors that is not true of mutualisms as defined above: if we consider a single individual performing the behavior in a group consisting solely of defectors, it would pay a cost without receiving any benefit. For example, if a single tent caterpillar starts to build a tent and the others save the energy they would have expended producing silk, then this focal individual pays all the cost and reaps only its small share of the benefit. If this cost outweighs the benefit, how then can such co-operation evolve?

For two-player interactions, this type of social dilemma can be formalized by setting the direct benefits of a co-operative behavior to be less than the costs, i.e., $D < C$. If the focal individual knows its partner will co-operate, then it is always better to co-operate, since as in mutualism, $B + E - C > B$. However, if the focal knows its partner will defect, then because $D < C$, it is better to also defect and avoid a negative payoff. In figure 10.1 this scenario is called synergism and has two possible evolutionarily stable states: one corresponding to everyone co-operating and another corresponding to everyone defecting. To which evolutionarily stable state the population evolves depends on the initial conditions. If the population initially contains more than $(C - D)/(E - D)$ co-operators, then evolution will lead to full co-operation, otherwise evolution will lead to full defection.

A prediction of the synergism model is that costly behavior can evolve even if, when interacting with other individuals that defect, the focal individual gets no benefit from co-operating. This point is not always given full consideration when discussing the evolution of costly signals. For example, cliff swallows call to signal the location of insect swarms thus paying a probably small cost in lost food, but providing nearby foraging partners a positive benefit in finding food. Brown et al. (1991) suggest, quite correctly, that swallows may have evolved call signaling because, "even if other birds do not also call, the caller could benefit through local enhancement simply by watching the nearby group members as some of them track the subsequent movement of the prey." If this is the case, then there may be no cost to interacting with a defector, i.e., $D > C$ and $E > C$, and full co-operation always evolves through mutualism. However, a benefit of local enhancement to signaling is not a requirement for the evolution of food calling. Rather, the game theory model predicts that provided there is an extra benefit when both birds call that is greater than the cost of calling, then co-operation can evolve independent of any direct benefits in the absence of calling, i.e., $D < C$. It is plausible that such extra benefits exist for cliff swallows. Groups that contain individuals that always signal can continuously track the movement of insect swarms. When interacting with a co-operator, the focal individual gets the additional benefit, $E > C$, of being able to re-find its own discovery. Defection would reduce both the focal and the partner bird's ability to find food.

More than the Sum of Its Parts

The last paragraph takes a two-player game and suggests it may apply to multi-player interactions. Swallows do not forage in pairs but rather in large groups. Under what circumstances can synergistic co-operation persist in larger groups? Box 10.C describes a continuous strategy game with group size N in which each individual can make an investment in co-operation. This investment incurs a constant cost, but group productivity increases as a function of the co-operative investment. This productivity is shared equally among individuals, so benefit to an individual increases as productivity divided by number of group members. Co-operation is evolutionarily stable for large groups in this model provided that group productivity increases with at least the square of the group size, or equivalently provided that the benefit per individual increases at least linearly with group size.

Figures 10.3 and 10.4 show how the evolutionarily stable states change with group size for two different productivity functions. When productivity grows superlinearly with group size then maximum co-operation is always a potential evolutionary outcome (figure 10.3). When productivity grows first superlinearly but then saturates, then maximum co-operation

Box 10.C Synergy in Groups of Size N

Consider a population that on each generation randomly aggregates in isolated groups of size N. Each individual can choose to invest an amount $p_i \in [0,1]$ in a co-operative behavior. The benefit to each individual, $g(\sum_{j=1}^{N} p_j)/N$, is assumed to be a function of the overall productivity of the group members, g, divided by the total number of group members. Assume that this function is the same for all group members. Thus the payoff for an individual i is

$$\frac{g(\sum_{j=1}^{N} p_j)}{N} - p_i c,$$

where c is the cost of the co-operative behavior. This model is an example of a structured-deme model (Nunney 1985; Wilson 1983).

Let us start by assuming that productivity increases as some power α of the level of co-operation of, i.e., $g(P) = bP^\alpha$. We now follow the method outlined by Doebeli et al. (2004). Assume that all individuals have the same strategy q apart from a mutant with strategy p. The selection gradient is then

$$D(q) = \frac{\partial}{\partial p} \left(\frac{b}{N} (p + (N-1)q)^\alpha - cp \right) \Bigg|_{p=q} = \frac{b\alpha}{N} (Nq)^{\alpha-1} - c.$$

Since we insist that investment is between 0 and 1, we can evaluate the selection gradient at these two extremes in order to see whether they are stable strategies. Evaluating $D(0) = -c$ tells us that the all defect is an evolutionarily stable state. Similarly, $D(1) = b\alpha N^{\alpha-2} - c$ tells us that the all co-operate is also evolutionarily stable, provided $b\alpha N^{\alpha-2} > c$. When all co-operate is stable there exists, although we do not determine it explicitly here, a single steady state q_* between these two extremes that is not convergent stable. This steady state acts as a repellant: when initially $q > q_*$ then $q \to 1$ and when initially $q < q_*$ then $q \to 0$. Qualitatively, the situation is the same as in the two-player discrete game discussed in the text: both all co-operate and all defect are evolutionarily stable.

The condition for synergistic co-operation in this model is $b\alpha N^{\alpha-2} > c$. If $\alpha < 2$ then as group size increases the cost an individual is willing to pay in co-operating decreases. For example, when $\alpha = 1$

we recover $b/N > c$. If $\alpha \geq 2$, however, then as group size increases the cost an individual is willing to pay tends toward a positive but finite limit. In particular, when $\alpha = 2$, co-operation is stable if $2b > c$ independent of N. Figure 10.3 shows how the steady states change with group size for $\alpha = 3$.

Figure 10.4 shows similar analysis for $g(P) = T^2P^3/(T^2 + P^2)$. This productivity function initially grows cubically, but when group size exceeds T the growth becomes more linear. For large P growth is purely linear. Here there are three different parameter regimes. For very small group sizes all individuals evolve to invest nothing ($p = 0$) in co-operation, but as group size increases the strategy of full investment ($p = 1$) becomes stable. At intermediate group sizes the full investment becomes unstable and a compromise of partial investment becomes stable. As group size increases still further all communication becomes evolutionarily unstable and $p = 0$ is the only evolutionarily stable state.

is stable for an intermediate range of group sizes (figure 10.4). This last observation is important, since realistic productivity functions must saturate at some point. Group productivity cannot continue to increase indefinitely with group size since at some point members of the large group must compete for resources (Foster 2004). Despite these diminishing returns for very large groups, co-operation is stable in intermediate-sized groups. This observation is true generally for productivity functions that increase superlinearly with group size at first and then saturate later (Sumpter & Brännström 2008).

Examples discussed in earlier chapters such as cliff swallow foraging, tent building by caterpillars and ant foraging involve either a group productivity that increases superlinearly with group size or, equivalently, a benefit per individual increasing linearly with group size. A key property of many synergistic interactions is the use of signals to spread information. Cliff sparrows use vocal signals (Brown et al. 1991); Norway rats (Galef & Buckley 1996), naked mole rats (Judd & Sherman 1996), and ants deposit residual trails; social insects use an array of different types of dances and other signals. The evolution of these signals is intimately linked with positive feedback. Signaling by a focal individual improves other group members' chances of discovering food; and since these group members are also signalers, then this improves the chance of rediscovering the same food or finding other nearby sources. The positive feedback continues and group productivity increases as more than the sum of the

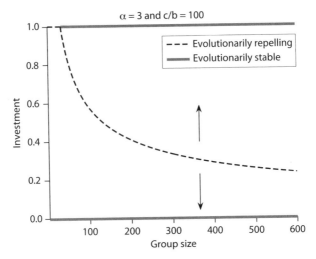

Figure 10.3. Model of synergy described in box 10.B with productivity that increases with the cube of group size, i.e., $g(P) = bP^3$. Bifurcation plots showing the location and stability of interior singular strategies and boundary points as a function of group size N. We choose $c/b = 100$ so that for very small groups there is no benefit to co-operation, i.e., $p = 0$ is the only stable strategy. With increasing group size a repelling interior singular strategy emerges and both no investment ($p = 0$) and maximal investment ($p = 1$) are locally stable strategies. Arrows indicate for which initial investment in co-operation these strategies will evolve (see (Sumpter & Brännström 2008) for details).

group's parts (see chapter 3). Such group productivity is precisely that needed for co-operation to evolve in the model in box 10.C.

Positive feedback and self-organization are sometimes proposed as alternatives to natural selection in explaining the evolution of co-operation. The argument is that because self-organized systems are more than the sum of their parts they cannot be understood simply in terms of the selfish individual units of which they are composed. By implication, the group becomes the unit upon which selection acts, and we need to consider how evolution functions on multiple levels (Fletcher & Doebeli 2006; Fletcher et al. 2006). The synergistic model I propose here shows that self-organization is consistent with selection on the level of the selfish individuals. Positive feedback can generate synergistic effects, and synergisms benefit the individual who is part of the group. The selection pressure on the individual is to co-operate, since it benefits from group membership. While the mechanisms whereby synergism is generated are often complex and need disentangling, this does not imply that their evolution cannot be understood at the level of the individual parts.

Synergism depends crucially on frequency dependence: co-operation can persist in an established population of co-operators, but cannot

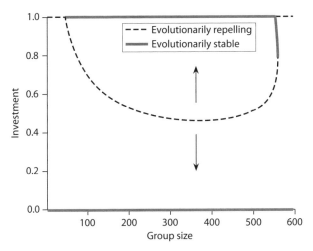

Figure 10.4. Model of synergy described in box 10.B with group productivity that first increases with the cube of group size but later saturates to linear increase, i.e., $g(P) = bT^2$ $P^3/(T^2 + P^2)$. Parameters are $c = 5$ and $T = 40$. Bifurcation plot showing the location and stability of interior singular strategies and boundary points as a function of group size N. As in fig. 10.3, with increasing group size a repelling interior singular strategy emerges and both no investment $(p = 0)$ and maximal investment $(p = 1)$ are locally stable strategies. In this case, however, as group sizes increases further $p = 1$ becomes unstable and a strategy corresponding to an intermediate investment in communication becomes stable. As group size increases still further the intermediate investment strategy disappears and $p = 0$ is the only stable state. The arrows from points indicate for which initial investment in co-operation the various stable strategies will evolve (see Sumpter & Brännström 2008 for details).

establish itself in an already established group of defectors. This is not always explicitly spelled out when co-operation is discussed. For example, West et al. (2006, 2007) classify social behaviors in a table with effect on actor in rows and effect on recipient in the columns. West et al. (2007) go on to try to clear up confusion about co-operation and altruism arising from inclusive fitness (see Inclusive Fitness in this chapter). However, in their classification they do not differentiate between direct and extra benefits and implicitly assume that $E = D$. Instead, they extend the model in box 10.A by allowing B to be negative (West et al. 2007, table 2, and for the more general case $N > 2$, table 3). Figure 10.5 interprets this classification in terms of the evolutionary game theory model in box 10.A. In contrast to figure 10.1, B is now on the x-axis instead of D, but B has no effect on evolutionary outcome. Equating E and D means that parasitism and synergism cannot occur, and the only possible evolutionarily stable states are those where either all individuals co-operate or all defect. Conversely, in the model in figure 10.1, when $E > C$ co-operation can evolve without altruism, even when $C > D$ (negative effect of co-operation on

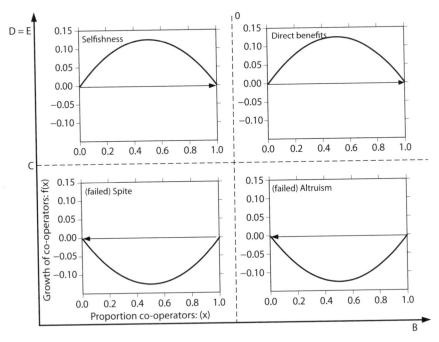

Figure 10.5. Classification of social behaviors for the two-player game proposed by West et al. (2006, 2007). Their model assumes that there is a cost to co-operating, $C > 0$, there is a direct benefit to self of co-operating independent of whether the other player defects, $D = E > 0$, and that a co-operator provides a benefit B to another individual. In the model given in table 2 of West et al. (2007) they further assume that $B = D$, but in the general model for types of co-operation given implicitly in table 1 of West et al. (2007) they allow B to have any value, positive or negative. Given these assumptions we can substitute these parameters into equation 10.A.1 in box 10.A. The figure shows how the cost benefit parameters, C, B, and D, determine the evolutionarily stable strategies. As before, B plays no role in determining the evolutionarily stable strategy.

actor at low, but not high, frequencies of co-operators) and $B > 0$ (positive effect on recipient at all frequencies).

To be fair to West et al., their classification is entirely valid and was designed to make the distinction between direct benefits and altruism. However, it is usually confusion between extra and direct benefits, and not misunderstanding of inclusive fitness, that leads to erroneous claims that "altruism" can arise from group selection or self-organization (Queller 1985; Queller & Strassmann 2006). Altruism is never predicted to occur in models with the assumptions I gave at the start of this chapter, while synergism can occur given a high enough initial proportion of co-operators.

One last point should be made about synergism. The model in box 10.C depends on a "fair" division of the payoffs. The fact that per capita productivity increases linearly with the number of individuals does not necessarily mean that the evolution of a particular form of co-operation is due to synergism. A central assumption that must also be tested is whether the benefits are on average shared equally among group members. If one individual has an opportunity to take consistently more than its share at the expense of the others then co-operation can fail. Unequal shares clearly occur in social insect colonies, where it is the small number of queens who produce the majority of offspring. In this case we need the additional explanation, in the form of increased inclusive fitness or some form of transactional concession to maintain co-operation.

Repeated Interactions

When there is a positive probability of interacting with the same individual again it can be beneficial to an individual to adopt a strategy that allows for co-operation. This observation is formalized in what is known as the iterated prisoner's dilemma (Axelrod & Hamilton 1981; Trivers 1971). In the iterated prisoner's dilemma, each individual plays a game a fixed number of times with payoffs as in table 10.1 with $D = E = 0$ and C and B positive. Over a wide range of conditions, a strategy known as Tit-for-Tat is evolutionarily stable for this game. A focal Tit-for-Tat individual co-operates on the first interaction and then on the next interaction adopts the same strategy as its partner did on the previous interaction. So if the partner defects, so too does the focal Tit-For-Tat individual. If two Tit-for-Tat individuals meet each other then they always co-operate. This interaction allows co-operation to evolve under the "threat" that defection by one individual will result in a break-off of co-operation and both individuals losing out.

There are many extensions to the iterated prisoner's dilemma, but in order for individuals to co-operate in these there must exist either a positive probability of interacting with the same individual (e.g., Killingback & Doebeli 2002; Schuessler 1989; Wahl & Nowak 1999) or indirect knowledge of a partner's previous strategic choices (Nowak & Sigmund 1998). Lehmann & Keller (2006a) provide a strong argument that, for all such models, a Tit-for-Tat strategy is evolutionarily stable provided that

$$mB > C,$$

where m is the probability that a further interaction will occur with the same individual multiplied by the probability that the focal individual

TABLE 10.2
Payoff table for iterated prisoners dilemma: Tit-for-Tat vs. All Defect

Focal / Partner	Tit-for-Tat	Always Defect
Tit-for-Tat	$(B - C)/(1 - m)$	$-C$
Always Defect	B	0

The Tit-for-Tat strategy co-operates on the first interaction and then copies its opponent's previous strategy (co-operate or defect) on further interactions. Always Defect strategy always defects. If a Tit-for-Tat meets an Always Defect individual it pays cost $-C$ on the first interaction, but then will itself always defect. The Always Defect individual gets benefit B on that first interaction, but then receives no further payoff. When Tit-for-Tats meet they continuously co-operate, both receiving payoff $B - C$ until their interaction breaks off. Since m is the probability that the interaction breaks off or a mistake in memory is made, the average number of co-operative interactions is $\sum_{i=1}^{\infty} i m^{i-1}(1 - m) = 1/(1 - m)$. Hence the average payoff is $(B - C)/(1 - m)$.

knows the result of the partner's last interaction. For this case, the payoff table for Tit-for-Tat playing against an Always Defect strategy is given in table 10.2. These payoffs are the same as those for synergistic interactions as in table 10.1, with $E = m(B - C)/(1 - m)$ and $D = 0$. The iterated prisoner's dilemma thus leads to either everyone being Tit-for-Tat or everyone being Always Defect, depending on the initial proportion of each in the population. A more complete description of strategies in the repeated prisoner's dilemma, including analysis of other potential alternative strategies, can be found in Hofbauer & Sigmund (1998).

There is some evidence that previous interactions and reputations are important in human interactions (Fehr & Fischbacher 2003, 2004; Skryms 2004) and in low cost activities by animals such as grooming and feeding. However, these have not been convincingly demonstrated in high cost co-operative breeding (Clutton-Brock 2002). Furthermore, unless there are strong preferential interactions within groups, we would expect that $m \propto 1/N$ and as group size increases the chance of repeated interactions will decrease. Unlike synergy, co-operation through repeated interactions relies on either small group sizes or a large capacity for remembering previous interactions. These observations have led several authors to question the general significance of repeated interactions alone in explaining the evolution of co-operation in many of the mass collective behavior's of animals (Clutton-Brock 2002; Richner & Heeb 1996). Indeed, repeated interactions do not play a significant role in explaining most of the collective behaviors I have discussed in this book. This is simply because the most interesting collective behaviors involve large numbers of individuals.

There are however various ways in which repeated interactions can enhance synergism or mutualism. For example, individuals with a choice of which group to join could first interact with group members to ascertain their behavioral strategy and then choose to join groups of co-operators rather than defectors. If payoff increases with group size then attracting other group members would provide an extra incentive for group members to co-operate with potential joiners (Kokko et al. 2002; Wilson & Dugatkin 1997). It is, however, difficult to envisage how such assortative interactions can lead to co-operation without the possibility of further repeated interactions or some form of synergism, mutualism, or parasitism. For example, let us assume co-operators can evolve a mechanism for identifying other co-operators and preferentially interact with them. But then what is to stop defectors evolving the same mechanism and exploiting co-operators? The only answer is to insist that the mechanism for identifying the co-operative feature is only found in those individuals that carry the feature (Dawkins 1976). While this remains a possibility, it is not a particularly general explanation of co-operation (Lehmann & Keller 2006a; West et al. 2006).

Inclusive Fitness

Altruism is a behavior that increases the direct fitness, or average lifetime reproductive success, of another individual while decreasing the direct fitness of the actor (West et al. 2006). In the models discussed so far, direct fitness is proportional to the payoff or fecundity of an individual minus the average payoff of the population (box 10.A). This definition automatically excludes evolutionarily stable strategies from being altruistic. If an individual can increase its direct fitness by changing strategy then a state is not evolutionarily stable. Altruism cannot evolve in such a setting (figure 10.1).

Hamilton's Rule

Hamilton proposed a simple yet far-reaching rule for the evolution of altruism. Assume that p and q are the respective behavioral strategies of a focal individual and its partner. Larger values of p and q can be thought of as corresponding to more investment in a co-operative behavior. Hamilton argued that if r is the co-efficient for relatedness between the focal individual and its partner, $b(p,q)$ is the total increase in direct fitness conferred on the partner by the focal individual and $c(p,q)$ is the cost of the behavior to the actor in terms of direct fitness, then the focal individual's investment in helping will increase if the *inclusive fitness* of the focal individual,

$$rb(p,q) - c(p,q),$$

is greater than zero. Inclusive fitness includes both the benefits of interaction that an individual gains directly in its own fitness plus those it gains indirectly through increased fitness of its relatives. The key idea underlying inclusive fitness is that, because related individuals have a higher probability of sharing the same genotype, genotypes with higher inclusive fitness are more likely to be transmitted to the next generation.

By interpreting p and q as the focal individual's and its partner's investment in co-operation, we can calculate the benefits and costs for interactions according to the payoffs in table 10.1. Using a method proposed by Taylor and Frank (1996) we find

$$b(p,q) = pE - pD + B \text{ and } c(p,q) = C - qE - (1-q)D$$

for evolutionary game theory models. Assuming that $D = E = 0$, then Hamilton's rule for an increase in helping is simply

$$rB > C. \tag{10.1}$$

If this inequality holds, a focal individual's probability of helping p will tend to 1, independent of the strategy of its partner. From the viewpoint of direct fitness this is altruism. When everyone co-operates, $q = 1$, a focal individual could increase its direct fitness by defecting. In doing so, however, it would reduce its inclusive fitness. Hamilton's insight was that in Darwinian natural selection, where the gene is the unit on which selection acts, it is inclusive fitness and not payoff, fecundity, or direct fitness that is maximized.

The relationship in equation 10.1 is a very specific version of Hamilton's rule. In the rule's more general form, the benefit $b(p,q)$ and cost $c(p,q)$ are functions of the strategies of the focal individual and the average strategy in the population. For example, if we relax the assumption that $D = E = 0$ then Hamilton's rule is

$$r(pE - pD + B) > C - qE - (1-q)D,$$

and a whole range of evolutionarily stable states for interactions between relatives arises. The rule now depends on the frequency of co-operators in the population. In general, we see that relatedness among individuals leads to an increase in co-operation: parameter combinations that would not have led to co-operation now move towards parasitism and synergy; and instances of parasitism and synergy become mutualisms (see table 10.3 for a summary of the effect of inclusive fitness).

Hamilton's rule has proven extremely successful in making predictions about co-operation and conflict in animal societies where between-individual relatedness is positive (Griffin & West 2003). Helping relatives

TABLE 10.3
Summary of results of model in box 10.B

	Parameters	Game Theory (figure 10.1)	Altruism	Repeated Interactions
Parasitism	$D > C > E$	$x_* = (C - D)/(E - D)$	x_* increases with r	x_* increases with m
Mutualism	$D > C, E > C$	$x_* = 1$	$x_* = 1$	$x_* = 1$
Synergism	$E > C > D$	$x_* = 0$ or $x_* = 1$	increased region of attraction to x_*	increased region of attraction to x_*
Altruism	$C > D, C > E$	$x_* = 0$	$x_* = 1$ if $rB > C$ (assuming $D = E = 0$)	$x_* = 0$, and $x_* = 1$ if $mB > C$

at a direct cost to oneself is a feature of all levels of biological organization: from bacteria to humans (Keller 1999). One of its most successful and intellectually interesting applications has been in understanding the evolution of insect societies (Bourke & Franks 1995; Queller & Strassmann 1998; Ratnieks et al. 2006). At the most basic level, the division of the colony into queens and workers is evidence of altruism. Nonreproducing workers help their mothers raise sisters who become queens and brothers who mate with queens from other colonies, thus passing their genes on to the next generation. A consistent failure to reproduce by a class of individuals is something that simply cannot evolve through natural selection in the absence of the gains in inclusive fitness implied by Hamilton's rule. Once the worker class is established, natural selection acts to increase the efficiency of the workers' interactions as colonies compete to produce the most queens and drones. It is here that the sophisticated forms of co-operation seen in honeybees, ants, and termites are seen to evolve (chapter 9).

Although the members of insect societies are usually related, the relatedness within these groups is often lower than predicted by their supposed family structure (Heinze et al. 2001; Korb & Heinze 2004). These observations have even led some to question whether relatedness has any importance at all in explaining co-operation (Costa 2006; Wilson & Holldobler 2005). At an extreme, colonies of unicolonial ant species, such as the Argentine ant, contain many unrelated queens (Pedersen et al. 2006). The question is how co-operation persists with such low levels or zero relatedness between group members? One explanation is a combination of synergism and altruism through indirect fitness benefits. Workers in insect societies gain inclusive fitness from their mother queens who in turn, even if unrelated to other queens, have an incentive to produce more workers and increase group efficiency. Greater insight into these questions could be gained by determining how colony performance increases with group size. Most importantly, synergistic interactions and helping of relatives should not be viewed as distinct explanations of co-operation, but different parts of a common explanation (Foster et al. 2006; West et al. 2007).

Family Groups and Spatially Structured Populations

The correlation between relatedness and co-operation is far from the complete story of Hamilton's rule or the application of inclusive fitness theory. Depending on the assumed lifecycle of individuals the predictions of inclusive fitness theory change. The derivation of equation 10.1 is based on the assumptions I made at the start of the chapter about the lifecycle of individuals. A question then arises as to where the between-individual relatedness originates from in such a model? At the start of this chapter, I assumed that individuals disperse widely and interact with others chosen randomly from the whole population. In the previous section I assumed that they interact more often than not with relatives. If these assumptions are not to contradict each other they need some clarification.

One way in which interactions can take place between relatives while simultaneously allowing global dispersal is to assume that groups initially consist of a single foundress, which then reproduces to found a "family" group of size N. We can then consider the interactions between the offspring of the foundress, who are positively related, and determine how much they should invest in co-operation. This approach can be further generalized by having n foundresses, or foundresses who have mated with n different males, where n is a lot smaller than the group size of offspring N. Multiple foundresses will reduce average relatedness in proportion to $1/n$ but relatedness will still have a positive effect on co-operation. One or a small number of foundresses building a larger nest of related offspring is common to the lifecycle of highly co-operative insect societies (Bourke & Franks 1995) and of bird species (Emlen 1997; Komdeur & Hatchwell 1999). Social insects (Downs & Ratnieks 1999; Wilson 1971) and birds (Sharp et al. 2005) have evolved mechanisms that ensure that they interact primarily with relatives. It is often this family structure that leads to altruistic interactions in these groups.

Without the introduction of family-based population structure, global dispersal means that within-group relatedness decreases to zero. On the other hand, local dispersal of individuals will always lead to some level of positive relatedness between neighbors. For example, assume that each group is formed on an island. At the end of each generation all the adults die and produce offspring, some of which remain on the island and others of which disperse to a randomly chosen island. There is a positive probability the individuals remaining on the same island share the same mother. This assumption is the basis of a mathematical model of local dispersal known as the continuous island model (Rousset 2004). Another way to introduce local dispersal or spatial population structure is to consider a ring of islands, each linked to one nearest neighbor. Dispersal then occurs between neighboring islands, and nearby individuals become more related than those living at far away islands. This spatial structure is an example of a stepping stone model (Kimura & Weiss 1964; Rousset 2004; Wright 1943).

Although local dispersal increases relatedness, it does not necessarily increase co-operation. Taylor (1992a, 1992b) showed that in both the island model and the stepping stone models the predicted level of co-operation is entirely independent of migration. This is because as migration decreases, not only does relatedness increase but so too does competition between relatives. In terms of Hamilton's rule, migration effects not only r but also $c(p,q)$ and $b(p,q)$, and it effects them in such a way that the inclusive fitness $r\,b(p,q) - c(p,q)$ remains constant (Rousset 2004). These theoretical ideas have been tested experimentally in fig wasps, where fighting between males was uncorrelated with their relatedness, but was negatively correlated with the level of competition they faced for future mating opportunities (West et al. 2001). In general, Taylor's result has deep implications for how relatedness is used to predict altruism. It shows that Hamilton's rule does not immediately imply that co-operation increases with relatedness.

While local migration alone does not necessarily promote co-operation, other details of the population's lifecycle do. Inclusive fitness theory predicts the conditions for co-operation over a wide range of assumptions about the lifecycle, including details of overlapping generations, and different forms of migration and niche construction (Irwin & Taylor 2000, 2001; Lehmann 2007; Taylor & Irwin 2000). In these models the inclusive fitness provides the condition under which co-operation can evolve (Grafen 1985; Lehmann & Keller 2006a; Rousset 2004). This breadth of application of inclusive fitness is sometimes underappreciated, probably because the simplicity of the inclusive fitness equation is deceiving. For any lifecycle the components r, $b(p,q)$, and $c(p,q)$ must all be calculated, and although there are extensive tools from population genetics that allow this to be done (Rousset 2004), it is not always straightforward to apply them.

A failure to see the proper connection to inclusive fitness has led to a confusing literature on how spatially local interactions can promote altruism (Lehmann & Keller 2006a, 2006b). Various simulation models, collectively known as evolutionary graph theory, have been proposed that show how local interactions can lead to altruism (Hauert 2002; Lieberman et al. 2005; Nowak & May 1992; Ohtsuki et al. 2006; Santos & Pacheco 2006). In these models, individuals play a prisoner's dilemma-like (i.e., D and E both less than C) game with local neighbors on a graph or social network. Depending on the form of the network and the details of the interaction rules, individuals evolve to have a greater or lesser tendency to "co-operate." These models do not usually follow the inclusive fitness approach of calculating relatedness between individuals. This proves problematic, since local migration always implies a positive relatedness. For example, Lehmann et al. (2007) showed that some basic results from evolutionary graph theory are special cases of the application of inclusive fitness theory. Altruism in many of these models arises simply from increased interactions with relatives, and not because spatially local interactions are a novel route to co-operation.

Although inclusive fitness theory provides some powerful tools for disentangling the evolution of co-operation in spatially structured populations, it is important to recognize that conditions promoting co-operation in particular biological and social systems cannot be derived without a good understanding of the details of the systems' life cycle (Ratnieks 2006). Models of the evolution of co-operation are often based on assumptions that adults produce large numbers of offspring, juveniles disperse independently of each other, and that there are the same number of individuals in all groups. Changes to these assumptions lead to changes in the underlying population structure and changes in predictions about the direction of selection. A multitude of different theoretical studies have investigated how co-operation evolves under different assumptions about population structure (e.g., Taylor 1992a, 1992b; Taylor and Frank 1996; Frank 1998; Taylor and Irwin 2000; Irwin and Taylor 2001; van Ballen & Rand 1998; Lehmann et al. 2006). This rich variety of theoretical studies reveals that there is no simple mathematical formula for the evolution of altruism, but that the inclusive fitness framework provides a powerful tool set to disentangle the factors that promote co-operation.

Unifying Explanations?

In the literature on social or group behavior there is no clear agreement on the exact definition of terms such as co-operation, social behavior, helping, mutual benefit, direct benefit, group selection, or even altruism. Different words have been defined by different authors to mean the same thing, and the same words have been defined to mean different things (West et al. 2006, 2007). An advantage of classifying co-operation as the outcome of different parameter values in a single model, as I have done here, is that the assumptions are clearly stated and the predictions are seen to be logical consequences of the model. I have of course chosen words to label different sets of assumptions and predictions and in doing so I chose mostly to follow (Clutton-Brock 2002), although instead of using his label of "group augmentation" (Kokko et al. 2001) I use Maynard Smith & Szathmáry's (1995) label synergism. However, labeling is not important. While these labels can change, the outcome of these two-player, two-strategy games will not. Given the same set of assumptions these models will make the same predictions.

In addition to analyzing two-player, two-strategy games, I have argued that many more complex game theory and inclusive fitness models can be classified under the same broad headings—parasitism, mutualism, synergism, repeated interactions, and altruism—that I have given the simpler models. There remain a large number of detailed models of animal co-operation that I have not covered here. These introduce a wide variety of concepts such as rent, punishment, partner choice, diminishing returns, etc. Each of these models is useful in describing particular situations and reveals how details of lifecycle produce changes in co-operation. I would, however, argue that in a broad sense they can be usefully classified as belonging to one or, depending on parameter values, a combination of the five headings I use above. Such a classification allows comparison among different systems. For example, we can say things like, "starling foraging and vigilance are both examples of social parasitism, while cliff swallow foraging appears to be synergistic." We may have built game theory models that capture the details of each of these systems but the classification allows us to quickly summarize what we have found out about these systems.

There is one last distinction that is worth drawing in the study of collective animal behavior: that is between behaviors that are strictly co-operative and those that are co-ordinating. The distinction is based on whether the cost to the focal individual has evolved or is simply a by-product of the individual undertaking an activity that is otherwise beneficial to itself (Clutton-Brock 2002). This is the same distinction that is made between cues and signals (Maynard Smith & Harper 2005). Cues can co-ordinate activities, but they have not evolved as a result of a benefit to the individuals that make them (chapter 3). Signals will only evolve when animals co-operate, and co-ordination always relies on cues. The distinction makes it straightforward to, for example, rule out social parasitism as an explanation of group foraging involving signals. The distinction is, however, less straightforward for mutualisms, where it is often difficult to distinguish whether a behavior has evolved to provide mutual benefit or whether the benefit was simply serendipitous.

There is richness in biological interactions that can never be captured by broad classifications. Indeed, simply answering questions such as how do birds flock, how do fish avoid predators, how do ants build their foraging networks, and how do cockroaches choose where to live with respect to their position on figure 10.1 does not tell us much about these fascinating problems. What makes these problems interesting usually has little to do with a broad notion of why they evolved. So while I have ended this book with an explanation of the functional or "why" questions in relation to collective animal behavior, the intention is not to finally unify the earlier chapters. Rather the aim of this chapter has been to clarify how functional reasoning is applied to these systems. The hope is that, together with the mechanistic description earlier in the book, we can use the combination of these approaches to build a full understanding of collective animal behavior.

— Chapter 11 —

Conclusions

This book has progressed from coming together through information transfer, decision-making, moving together, synchronization, structures, and regulation to finally arrive at complicated interactions. This progression has taken on increasingly complex aspects of collective animal behavior. Each chapter has attempted to unify group behavior of different species in these different situations and explain similarities in the underlying function and mechanisms. I will close this book with a brief discussion of how I believe we should think of the science of collective animal behavior and suggest some future directions for research.

From Toys to Tools

This book grew from a review article I wrote three years ago on "principles of collective animal behaviour" (Sumpter 2006). There I outlined several guiding principles such as "positive feedback," "individual variation," etc., which underlie many aspects of collective animal behavior. These principles have again appeared at many points within this book.

Having a set of such principles is useful for grouping and categorizing ideas and more quickly understanding new systems. Here, I have chosen a slightly different grouping of ideas under the different chapter headings, concentrating more on similarities between systems in spatial or temporal organization. Again, these chapters provide a way of unifying our understanding of different systems.

Another way of unifying different systems is through mathematical modeling. One of the most remarkable features of the study of collective animal behavior is the applicability of mathematical models. This is made all the more remarkable when we consider that animals are not as simple as physical particles. Individuals vary and experiments always involve intrinsic variation and noise. Despite, and sometimes because of this variation, it is possible to use one model to make predictions about very different types of groups.

Unifying principles are not, however, the be all and end all of science. In the 2006 article and here, I advocate a pragmatic view to unifying theories of collective behavior. The study of collective animal behavior should proceed on a case-by-case basis. For each particular system, we should classify how individuals interact with each other and build mathematical models based on observations. In many cases, models of one system may be applicable to other systems and this can help us understand the underlying mechanisms. This similarity between models should not in itself become the driving force in our research.

Instead, I see mathematical models and different theoretical approaches as a tool set for understanding a wide variety of systems. No single mathematical model provides a unique correct way of describing all aspects of a particular system. Neither can we expect to apply the same model to all systems. The art of understanding the world is not in mastery of particular models but in the ability to recognize and exploit connections where they exist.

It is for this reason that neither mechanistic nor functional approaches to group behavior can claim precedence over the other. Both of these approaches have produced elegant models, which have been tested against experiments and suggest new ways in which systems can be viewed. In this book, I have emphasized the combination of functional and mechanistic approaches. Even if particular studies are likely to be more or less biased to either the functional or the mechanistic, it is important to bear in mind how the other approach can play a role in increasing understanding.

The idea that mathematical modeling has a role to play in understanding complex systems is not new. The last 30 years has seen the rapid growth of complexity science, the application of non-linear mathematics, statistical physics, and the theory of networks to understanding biochemistry, biology, and sociology. In many cases these models are "toy models" of systems. The Kuramoto model, self-propelled particles, the logistic equation, self-organized criticality, small world networks, voter models, preferential attachment, to name just a few, are ideas that are less inspired by details of particular systems and more an attempt to abstract from details and make general predictions about a wide range of systems.

In my opinion, the aim of complexity science today should be to move from these toy models toward a set of tools that can be applied to specific complex systems. Without a clear relationship to biological or sociological systems the role of toy models is limited. Or to put it more bluntly, as a biologist colleague once told me as I was trying to explain one such model to him, "Leave your toys at home, we're trying to work here."

The models applied in this book are not just toys used to train our intuition before we get to work on the real thing. They are the tools of a serious approach to understanding specific systems. From models of animal group size distributions; through cockroach, honeybee, and ant emigration; to the collective motion of insects, birds, and fish; to the structure of ant trails and termite nests, the interplay between experiment and model is clear. The models bring rigor to our assumptions about a system, make testable predictions, and in many cases provide a quantitative as well as a qualitative match to the available data.

It is the combination of the relative sophistication of the individuals of which animal groups are composed, and the empirical success in using mathematical models to predict and understand their behavior that makes the study of collective animal behavior important. Despite the seeming complexity of the task of understanding social interactions, we do have tools that allow us to predict experimental outcomes. This should give hope to applications of similar methods in understanding ecosystems, brain function, and other complex systems.

Unfortunately, for my biologist friend, there is no clear cut distinction between toys and tools. Often a toy model based on very little experimental insight can serve as a basis for a more rigorous and detailed argument when fleshed out. Indeed, models of animal aggregations started as toy models of particle cohesion, and self-propelled particle models of animal motion started as animations for computer games!

The important question is how to get the balance right between simplifying assumptions and biological detail. I believe the key to getting this balance right is as follows: one should concentrate on models that make predictions about a system that are both non-trivial and testable. Non-trivial means that the model is needed to make the prediction. We cannot simply arrive at the same conclusion by verbal argument alone. Here the rigor brought by mathematical modeling is important. Often the problem with verbal arguments is not that they cannot be used to make a particular prediction, but rather that verbal arguments can be made for all sorts of different predictions. It is determining which prediction follows from a set of well stated assumptions that is non-trivial and is the tool provided by mathematical modeling.

The work described in this book is testimony to the fact that ideas like self-organization, emergence, and complexity theory are not just fancy sounding names, but can be applied to make non-trivial and testable predictions about biological and sociological systems. For this reason, scientists in all fields should be interested in collective animal behavior and mathematical modeling of complex systems in general.

Some Open Questions

There are many open and interesting questions in collective animal behavior. I have mentioned many of these during the course of the book, but it is worth listing some of them here for further consideration.

Linking mechanisms and function in group size distribution (chapter 2). While Niwa's model appears to provide a good empirical fit to data on fish schools and may be extendable to other animal groups, it does not include functional considerations. An interesting question is how individuals manipulate the average group size they experience, the key parameter in Niwa's model.

How humans integrate many wrongs (chapters 3 & 4). The Milgram and Asch observations and experiments on humans reveal that humans use quorum-like rules to decide whether to copy the choices of others. It would be interesting to investigate these ideas in a context where there is a reward to be gained by making correct choices. Here, laboratory experiments, similar to those performed to investigate the prisoner's dilemma and other co-operative games in humans, could be used.

Modeling realistic motion of bird flocks and fish schools (chapter 5). Many of the "mesoscopic" features of moving animal groups are not reproduced by current self-propelled particle models. The main restriction here is the availability of empirical data on the structure and dynamics of moving animal groups. It is currently difficult to quantify why the models are not quite right. Recent empirical work on starlings is beginning to provide this data, but it remains an important challenge in an area where so much theoretical modeling has already been done.

Evolving self-propelled particles (chapter 5). The Wood et al. model of how predation avoidance evolves is just one example of how natural selection might influence collective motion. One possibility is that natural selection acts to increase the complexity of group motion, so that the group is highly sensitive to changes in the environment (Sumpter et al. 2008a).

Individual variation (chapters 4 & 6). Many of the models of collective behavior assume that individuals are identical units, but the many wrongs idea and Kuramoto's model of synchronization instead make predictions on the basis of differences among individuals. There are consistent differences among animals in their behavior and a research challenge is to understand how these differences are integrated at the level of the group. Instead of seeing individual differences as simply "noise" we should investigate their role in producing collective patterns.

Modeling complex nest structures (chapter 7). Although several models have explained formation of pillars and chambers via templates and stigmergy, how the complex structures such as harvester ant nests (figure 7.5) are constructed remains an open problem.

Providing a useable formal framework for individual-based modeling (chapter 9). One of the weaknesses of the individual-based approach to modeling is that individual-based model results can be difficult to reliably replicate because they are not expressed in a formal framework. The problem is agreeing on a tool for individual-based modeling and designing one that is flexible enough to encompass different types of models.

Collective human behavior (chapters 3, 6, 7, & 8). While there is a growing application of mathematical models to understand the social behavior of humans, I would continue to classify many of the models as toys rather than tools. The studies discussed throughout this book are notable exceptions and other recent studies using network analysis to look at social interaction, for example, on the internet are also beginning to use data to inform models. The possibility for rigorous application to data on human social interactions is clear. It will be interesting to see how the types of techniques used in studying collective animal behavior can be applied in studying humans.

References

Acosta, F. J., Lopez, F. & Serrano, J. M. 1993. Branching angles of ant trunk trails as an optimization cue. *Journal of Theoretical Biology* **160**, 297–310.

Adams, E. S. & Tschinkel, W. R. 1995. Density-dependent competition in fire ants—effects on colony survivorship and size variation. *Journal of Animal Ecology* **64**, 315–324.

Aleksiev, A. S., Sendova-Franks, A. B. & Franks, N. R. 2007. Nest "moulting" in the ant *Temnothorax albipennis*. *Animal Behaviour* **74**, 567–575.

Amé, J. M., Halloy, J., Rivault, C., Detrain, C. & Deneubourg, J. L. 2006. Collegial decision making based on social amplification leads to optimal group formation. *PNAS* **103**, 5835–5840.

Amé, J. M., Rivault, C. & Deneubourg, J. L. 2004. Cockroach aggregation based on strain odour recognition. *Animal Behaviour* **68**, 793–801.

Anderson, C. & Jadin, J. L. V. 2001. The adaptive benefit of leaf transfer in Atta colombica. *Insectes Sociaux* **48**, 404–405.

Anderson, C. & McShea, D. W. 2001a. Individual versus social complexity, with particular reference to ant colonies. *Biological Review* **76**, 211–237.

Anderson, C. & McShea, D. W. 2001b. Intermediate-level parts in insect societies: Adaptive structures that ants build away from the nest. *Insectes Soc.* **48**, 291–301.

Anderson, C. & Ratnieks, F. L. W. 1999. Worker allocation in insect societies: coordination of nectar foragers and nectar receivers in honey bee (*Apis mellifera*) colonies. *Behavioral Ecology and Sociobiology* **46**, 73–81.

Anderson, C. & Ratnieks, F. L. W. 2000. Task partitioning in insect societies: Novel situations. *Insectes Sociaux* **47**, 198–199.

Aoki, I. 1982. A simulation study on the schooling mechanism in fish. *Bulletin of the Japanese Society of Scientific Fisheries* **48**, 1081–1088.

Arthur, W. B. 1994. Inductive reasoning and bounded rationality. *American Economic Review* **84**, 406–411.

Arthur, W. B. 1999. Complexity and the economy. *Science* **284**, 107–109.

Asch, S. E. 1955. Opinions and social pressure. *Scientific American* **193**, 31–25.

Ashby, W. R. 1947. Principles of the self-organizing dynamic system. *Journal of General Psychology* 125–128.

Avery, C. & Zemsky, P. 1998. Multidimensional uncertainty and herd behavior in financial markets. *American Economic Review* **88**, 724–748.

Aviles, L. & Tufino, P. 1998. Colony size and individual fitness in the social spider *Anelosimus eximius*. *American Naturalist* **152**, 403–418.

Avitabile, A., Morse, R. A. & Boch, R. 1975. Swarming honey bees guided by pheromones. *Annals of the Entomological Society of America* **6**, 1079–1082.

Axelrod, R. & Hamilton, W. D. 1981. The Evolution of cooperation. *Science* **211**, 1390–1396.

Azcarate, F. M. & Peco, B. 2003. Spatial patterns of seed predation by harvester ants (*Messor Forel*) in Mediterranean grassland and scrubland. *Insectes Soc.* **50**, 120–126.

Bak, P. 1996. *How nature works: The science of self-organized criticality.* New York: Copernicus Books.

Ball, P. 2004. *Critical mass: How one thing leads to another.* London: Heinemann.

Ballerini, M., Cabibbo, N., Candelier, R., Cavagna, A., Cisbani, E., Giardina, I., Orlandi, A., Parisi, G., Procaccini, A., Viale, M. & Zdravkovic, V. 2008a. Empirical investigation of starling flocks: A benchmark study in collective animal behaviour. *Animal Behaviour* **76**, 201–215.

Ballerini, M., Cabibbo, N., Candeleir, R., Cavagna, A., Cisbani, E., Giardina, I., Lecomte, V., Orlandi, A., Parisi, G., Procaccini, A., Viale, M. & Zdravkovic, V. 2008b. Interaction ruling animal collective behavior depends on topological rather than metric distance: Evidence from a field study. *Proceedings of the National Academy of Sciences of the United States of America* **105**, 1232–1237.

Barabasi, A. L. 2003. *Linked.* London: Penguin Books.

Barabasi, A. L. & Albert, R. 1999. Emergence of scaling in random networks. *Science* **286**, 509–512.

Barabasi, A. L., Albert, R. & Jeong, H. 1999. Mean-field theory for scale-free random networks. *Physica A* **272**, 173–187.

Barber, J. C. E. 2001. Social influences on the motivation of laying hens. Thesis in *Department of Zoology*, vol. DPhil. Oxford: University of Oxford.

Barnard, C. J. & Sibly, R. M. 1981. Producers and scroungers—a general-model and its application to captive flocks of house sparrows. *Animal Behaviour* **29**, 543–550.

Bartholdi, J. J., Seeley, T. D., Tovey, C. A. & Vande Vate, J. H. 1993. The pattern and effectiveness of forager allocation among flower patches by honey bee colonies. *Journal of Theoretical Biology* **160**, 23–40.

Batty, M. 2008. The size, scale, and shape of cities. *Science* **319**, 769–771.

Batty, M., Desyllas, J. & Duxbury, E. 2003a. The discrete dynamics of small-scale spatial events: Agent-based models of mobility in carnivals and street parades. *International Journal of Geographical Information Science* **17**, 673–697.

Batty, M., Desyllas, J. & Duxbury, E. 2003b. Safety in numbers? Modeling crowds and designing control for the Notting Hill Carnival. *Urban Studies* **40**, 1573–1590.

Bays, P. M. & Wolpert, D. M. 2007. Computational principles of sensorimotor control that minimize uncertainty and variability. *Journal of Physiology–London* **578**, 387–396.

Beauchamp, G. & Fernandez-Juricic, E. 2005. The group-size paradox: Effects of learning and patch departure rules. *Behavioral Ecology* **16**, 352–357.

Becco, C. V., Vanewalle, N., Delcourt, J. & Poncin, P. 2006. Experimental evidences of a structural and dynamical transition in fish school. *Physica A* **367**, 487–493.

Beckers, R., Deneubourg, J. L. & Goss, S. 1992a. Trail laying behavior during food recruitment in the ant *Lasius niger* (L). *Insectes Sociaux* **39**, 59–72.

Beckers, R., Deneubourg, J. L. & Goss, S. 1992b. Trails and U-turns in the selection of a path by the ant *Lasius niger*. *Journal of Theoretical Biology* **159**, 397–415.

Beckers, R., Deneubourg, J. L. & Goss, S. 1993. Modulation of trail laying in the ant *Lasius niger* (Hymenoptera, Formicidae) and its role in the collective selection of a food source. *Journal of Insect Behavior* **6**, 751–759.

Beckers, R., Goss, S., Deneubourg, J. L. & Pasteels, J. M. 1989. Colony size, communication and ant foraging strategy. *Psyche* **96**, 239–256.

Bednekoff, P. A. 1997. Mutualism among safe, selfish sentinels: A dynamic game. *American Naturalist* **150**, 373–392.

Bednekoff, P. A. & Woolfenden, G. E. 2003. Florida scrub-jays (*Aphelocoma coerulescens*) are sentinels more when well-fed (even with no kin nearby). *Ethology* **109**, 895–903.

Bednekoff, P. A. & Woolfenden, G. E. 2006. Florida scrub-jays compensate for the sentinel behavior of flockmates. *Ethology* **112**, 796–800.

Beekman, M. & Bin Lew, J. 2008. Foraging in honeybees—when does it pay to dance? *Behavioral Ecology* **19**, 255–262.

Beekman, M., Fathke, R. L. & Seeley, T. D. 2006. How does an informed minority of scouts guide a honeybee swarm as it flies to its new home? *Animal Behaviour* **71**, 161–171.

Beekman, M., Gilchrist, A. L., Duncan, M. & Sumpter, D. J. T. 2007. What makes a honeybee scout? *Behavioral Ecology and Sociobiology* **61**, 985–995.

Beekman, M., Komdeur, J. & Ratnieks, F. L. W. 2003. Reproductive conflicts in social animals: Who has power? *Trends in Ecology & Evolution* **18**, 277–282.

Beekman, M., Sumpter, D. J. T. & Ratnieks, F. L. W. 2001. Phase transition between disordered and ordered foraging in Pharaoh's ants. *Proceedings of the National Academy of Sciences of the United States of America* **98**, 9703–9706.

Bennett, M., Schatz, M. F., Rockwood, H. & Wiesenfeld, K. 2002. Huygens's clocks. *Proceedings of the Royal Society of London Series a–Mathematical Physical and Engineering Sciences* **458**, 563–579.

Bernasconi, G., Ratnieks, F. L. W. & Rand, E. 2000. Effect of "spraying" by fighting honey bee queens (*Apis mellifera L.*) on the temporal structure of fights. *Insectes Sociaux* **47**, 21–26.

Bernhardt, D., Campello, M. & Kutsoati, E. 2006. Who herds? *Journal of Financial Economics* **80**, 657–675.

Bernstein, C., Kacelnik, A. & Krebs, J. R. 1988. Individual decisions and the distribution of predators in a patchy environment. *Journal of Animal Ecology* **57**, 1007–1026.

Berryman, A. A. 1999. *Principles of population dynamics*. London: Stanley Thornes.

Bertram, B. C. R. 1975. Social factors influencing reproduction in wild lions. *Journal of Zoology* **177**, 463–482.

Bettencourt, L. M. A., Lobo, J., Helbing, D., Kuhnert, C. & West, G. B. 2007. Growth, innovation, scaling, and the pace of life in cities. *Proceedings of the National Academy of Sciences of the United States of America* **104**, 7301–7306.

Biesmeijer, J. C. & de Vries, H. 2001. Exploration and exploitation of food sources by social insect colonies: A revision of the scout-recruit concept. *Behavioral Ecology and Sociobiology* **49**, 89–99.

Biesmeijer, J. C. & Slaa, E. J. 2004. Information flow and organization of stingless bee foraging. *Apidologie* **35**, 143–157.

Biro, D., Sumpter, D. J. T., Meade, J. & Guilford, T. 2006. From compromise to leadership in pigeon homing. *Current Biology* **16**, 2123–2128.

Boi, S., Couzin, I. D., Del Buono, N., Franks, N. R. & Britton, N. F. 1999. Coupled oscillators and activity waves in ant colonies. *Proceedings of the Royal Society of London Series B–Biological Sciences* **266**, 371–378.

Bollazzi, M. & Roces, F. 2007. To build or not to build: Circulating dry air organizes collective building for climate control in the leaf-cutting ant *Acromyrmex ambiguus*. *Animal Behaviour* **74**, 1349–1355.

Bonabeau, E. & Dagorn, L. 1995. Possible universality in the size distribution of fish schools. *Physical Review E* **51**, R5220–R5223.

Bonabeau, E., Dagorn, L. & Freon, P. 1999. Scaling in animal group-size distributions. *Proceedings of the National Academy of Sciences of The United States of America* **96**, 4472–4477.

Bonabeau, E., Theraulaz, G. & Deneubourg, J. L. 1998a. Fixed response thresholds and the regulation of division of labor in insect societies. *Bulletin of Mathematical Biology* **60**, 753–807.

Bonabeau, E., Theraulaz, G., Deneubourg, J. L., Aron, S. & Camazine, S. 1997. Self-organization in social insects. *Trends in Ecology & Evolution* **12**, 188–193.

Bonabeau, E., Theraulaz, G., Deneubourg, J. L., Franks, N. R., Rafelsberger, O., Joly, J. L. & Blanco, S. 1998b. A model for the emergence of pillars, walls and royal chambers in termite nests. *Philosophical Transactions of the Royal Society of London Series B–Mathematical Physical and Engineering Sciences* **353**, 1561–1576.

Bono, J. M. & Crespi, B. J. 2006. Costs and benefits of joint colony founding in Australian Acacia thrips. *Insectes Sociaux* **53**, 489–495.

Bono, J. M. & Crespi, B. J. 2008. Cofoundress relatedness and group productivity in colonies of social *Dunatothrips* (Insecta : Thysanoptera) on Australian *Acacia*. *Behavioral Ecology and Sociobiology* **62**, 1489–1498.

Bortkiewicz, L. J. 1898. *The Law of Small Numbers*. Leipzig: Teubner.

Bourke, A. F. G. & Franks, N. R. 1995. *Social evolution in ants*. Princeton, New Jersey: Princeton University Press.

Brannstrom, A. & Sumpter, D. J. T. 2005. The role of competition and clustering in population dynamics. *Proceedings of the Royal Society B–Biological Sciences* **272**, 2065–2072.

Britton, N. F. 2005. *Essential Mathematical Biology*. London: Springer.

Britton, N. F., Franks, N. R., Pratt, S. C. & Seeley, T. D. 2002. Deciding on a new home: How do honeybees agree? *Proceedings of the Royal Society of London Series B–Biological Sciences* **269**, 1383–1388.

Brown, C. R. 1986. Cliff swallow colonies as information-centers. *Science* **234**, 83–85.

Brown, C. R. & Brown, M. B. 1986. Ectoparasitism as a cost of coloniality in cliff swallows (*Hirundo pyrrhonota*). *Ecology* **67**, 1206–1218.

Brown, C. R. & Brown, M. B. 1996. *Coloniality in the cliff swallow*. Chicago: The University of Chicago Press.

Brown, C. R., Brown, M. B. & Shaffer, M. L. 1991. Food-sharing signals among socially foraging cliff swallows. *Animal Behaviour* **42**, 551–564.

Bruch, E. E. & Mare, R. D. 2006. Neighborhood choice and neighborhood change. *American Journal of Sociology* **112**, 667–709.

Bruns, G. 1997. *Distributed systems analysis with CCS*: Prentice Hall.

Buchanan, M. 2000. *Ubiquity: The new science that is changing the world*. London: Phoenix.

Buck, J. 1988. Synchronous rhythmic flashing of fireflies .2. *Quarterly Review of Biology* **63**, 265–289.

Buck, J. & Buck, E. 1976. Synchronous fireflies. *Scientific American* **234**, 74–85.

Buck, J., Buck, E., Case, J. F. & Hanson, F. E. 1981. Control of flashing in fireflies .5. Pacemaker synchronization in *Pteroptyx cribellata*. *Journal of Comparative Physiology* **144**, 287–298.

Buhl, J., Deneubourg, J. L., Grimal, A. & Theraulaz, G. 2005. Self-organized digging activity in ant colonies. *Behavioral Ecology and Sociobiology* **58**, 9–17.

Buhl, J., Gautrais, J., Deneubourg, J. L. & Theraulaz, G. 2004. Nest excavation in ants: Group size effects on the size and structure of tunneling networks. *Naturwissenschaften* **91**, 602–606.

Buhl, J., Hicks, K., Miller, E., Persey, S., Alinvi, O. & Sumpter, D. 2009. Shape and efficiency of wood ant foraging networks. *Behavioral Ecology and Sociobiology* **63**, 451–460.

Buhl, J., Sumpter, D. J. T., Couzin, I. D., Hale, J. J., Despland, E., Miller, E. R. & Simpson, S. J. 2006. From disorder to order in marching locusts. *Science* **312**, 1402–1406.

Burd, M. 2006. Ecological consequences of traffic organisation in ant societies. *Physica a–Statistical Mechanics and Its Applications* **372**, 124–131.

Burd, M. & Aranwela, N. 2003. Head-on encounter rates and walking speed of foragers in leaf-cutting ant traffic. *Insectes Sociaux* **50**, 3–8.

Burt, C. 1963. Is intelligence distributed normally? *British Journal of Mathematical & Statistical Psychology*, 170–190.

Calhim, S., Shi, J. B. & Dunbar, R. I. M. 2006. Sexual segregation among feral goats: Testing between alternative hypotheses. *Animal Behaviour* **72**, 31–41.

Camazine, S., Deneubourg, J. L., Franks, N. R., Sneyd, J., Theraulaz, G. & Bonabeau, E. 2001. *Self-organization in biological systems*. Princeton studies in complexity. Princeton, New Jersey: Princeton University Press.

Camazine, S. & Sneyd, J. 1991. A model of collective nectar source selection by honey bees: Self-organisation through simple rules. *Journal of Theoretical Biology* **149**, 547–571.

Camazine, S., Visscher, P. K., Finley, J. & Vetter, R. S. 1999. House-hunting by honey bee swarms: Collective decisions and individual behaviors. *Insectes Sociaux* **46**, 348–360.

Cammaerts, M. C. & Cammaerts, R. 1980. Food recruitment strategies of the ants *Myrmica sabuleti* and *Myrmica ruginodis*. *Behavioural Processes* **5**, 251–270.

Caraco, T. 1979a. Time budgeting and group-size—test of theory. *Ecology* **60**, 618–627.

Caraco, T. 1979b. Time budgeting and group-size—theory. *Ecology* **60**, 611–617.

Caraco, T., Martindale, S. & Pulliam, H. R. 1980. Avian flocking in the presence of a predator. *Nature* **285**, 400–401.

Carlson, J. M. & Doyle, J. 2002. Complexity and robustness. *Proceedings of the National Academy of Sciences of the United States of America* **99**, 2538–2545.

Cavagna, A., Cimarelli, A., Giardina, I., Orlandi, A., Parisi, G., Procaccini, A., Santagati, R. & Stefanini, F. 2008a. New statistical tools for analyzing the structure of animal groups. *Mathematical Biosciences* **214**, 32–37.

Cavagna, A., Giardina, I., Orlandi, A., Parisi, G., Procaccini, A., Viale, M. & Zdravkovic, V. 2008b. The STARFLAG handbook on collective animal behaviour: 1. Empirical methods. *Animal Behaviour* **76**, 217–236.

Challet, D. & Zhang, Y. C. 1997. Emergence of cooperation and organization in an evolutionary game. *Physica A* **246**, 407–418.

Challet, D. & Zhang, Y. C. 1998. On the minority game: Analytical and numerical studies. *Physica A* **256**, 514–532.

Charles, C. Z. 2003. The dynamics of racial residential segregation. *Annual Review of Sociology* **29**, 167–207.

Chauvin, R. 1962. Observations sur les pistes de *Formica polyctena*. *Insectes Soc.* **9**, 311–321.

Chretien, L. 1996. Organisation spatiele du matériel provenant de l'excavation du nid chez *Messor barbarus* et des cadavres d'ouvrières chez *Lasius niger* (Hymenoptera: Formicidae). In *center for nonlinear phenomena and complex systems* vol. PhD. Brussels: Université Libre de Bruxelles.

Clark, C. W. & Mangel, M. 1984. Foraging and flocking strategies—information in an uncertain environment. *American Naturalist* **123**, 626–641.

Clark, C. W. & Mangel, M. 1986. The evolutionary advantages of group foraging. *Theoretical Population Biology*, 45–75.

Clark, W. A. V. 1991. Residential preferences and neighborhood racial segregation—a test of the Schelling segregation model. *Demography* **28**, 1–19.

Clutton-Brock, T. 2002. Breeding together: Kin selection and mutualism in cooperative vertebrates. *Science* **296**, 69–72.

Clutton-Brock, T. H., Brotherton, P. N. M., Russell, A. F., O'Riain, M. J., Gaynor, D., Kansky, R., Griffin, A., Manser, M., Sharpe, L., McIlrath, G. M., Small, T., Moss, A. & Monfort, S. 2001. Cooperation, control, and concession in meerkat groups. *Science* **291**, 478–481.

Clutton-Brock, T. H., O'Riain, M. J., Brotherton, P. N. M., Gaynor, D., Kansky, R., Griffin, A. S. & Manser, M. 1999. Selfish sentinels in cooperative mammals. *Science* **284**, 1640–1644.

Cole, B. J. 1991a. Is animal behavior chaotic—evidence from the activity of ants. *Proceedings of the Royal Society of London Series B–Biological Sciences* **244**, 253–259.

Cole, B. J. 1991b. Short-term activity cycles in ants: Generation of periodicity through worker interaction. *American Naturalist* **137**, 244–259.

Cole, B. J. & Cheshire, D. 1996. Mobile cellular automata models of ant behavior: Movement activity of *Leptothorax allardycei*. *American Naturalist* **148**, 1–15.

Collett, M., Despland, E., Simpson, S. J. & Krakauer, D. C. 1998. Spatial scales of desert locust gregarization. *Proceedings Of The National Academy of Sciences of the United States of America* **95**, 13052–13055.

Collins, L. M. & Sumpter, D. J. T. 2007. The feeding dynamics of broiler chickens. *Journal of the Royal Society Interface* **4**, 65–72.

Conlisk, J. 1996. Why bounded rationality? *Journal of Economic Literature* **34**, 669–700.

Conradt, L. 1998. Could asynchrony in activity between the sexes cause intersexual social segregation in ruminants? *Proceedings of the Royal Society of London Series B–Biological Sciences* **265**, 1359–1363.

Conradt, L. & Roper, T. J. 2000. Activity synchrony and social cohesion: A fission-fusion model. *Proceedings of the Royal Society of London Series B–Biological Sciences* **267**, 2213–2218.

Conradt, L. & Roper, T. J. 2003. Group decision-making in animals. *Nature* **421**, 155–158.

Conradt, L. & Roper, T. J. 2005. Consensus decision making in animals. *Trends In Ecology & Evolution* **20**, 449–456.

Conradt, L. & Roper, T. J. 2007. Democracy in animals: The evolution of shared group decisions. *Proceedings of the Royal Society B–Biological Sciences* **274**, 2317–2326.

Coolen, I., Giraldeau, L. A. & Lavoie, M. 2001. Head position as an indicator of producer and scrounger tactics in a ground-feeding bird. *Animal Behaviour* **61**, 895–903.

Costa, J. T. 2006. *The other insect societies*. Cambridge, Massachusetts: Harvard University Press.

Costa, J. T. & Ross, K. G. 2003. Fitness effects of group merging in a social insect. *Proceedings of the Royal Society B–Biological Sciences* **270**, 1697–1702.

Couzin, I. D. & Franks, N. R. 2003. Self-organized lane formation and optimized traffic flow in army ants. *Proceedings of The Royal Society of London Series B–Biological Sciences* **270**, 139–146.

Couzin, I. D. & Krause, J. 2003. Self-organization and collective behavior in vertebrates. *Advances in the Study of Behavior* **32**, 1–75.

Couzin, I. D., Krause, J., Franks, N. R. & Levin, S. A. 2005. Effective leadership and decision-making in animal groups on the move. *Nature* **433**, 513–516.

Couzin, I. D., Krause, J., James, R., Ruxton, G. D. & Franks, N. R. 2002. Collective memory and spatial sorting in animal groups. *Journal of Theoretical Biology* 218, 1–11.

Croft, D. P., Arrowsmith, B. J., Bielby, J., Skinner, K., White, E., Couzin, I. D., Magurran, A. E., Ramnarine, I. & Krause, J. 2003. Mechanisms underlying shoal composition in the Trinidadian guppy, *Poecilia reticulata*. *Oikos* 100, 429–438.

Croft, D. P., James, R., Thomas, P. O. R., Hathaway, C., Mawdsley, D., Laland, K. N. & Krause, J. 2006. Social structure and co-operative interactions in a wild population of guppies (*Poecilia reticulata*). *Behavioral Ecology and Sociobiology* 59, 644–650.

Croft, D. P., James, R., Ward, A. J. W., Botham, M. S., Mawdsley, D. & Krause, J. 2005. Assortative interactions and social networks in fish. *Oecologia* 143, 211–219.

Cully, S. M. & Seeley, T. D. 2004. Self-assemblage formation in a social insect: The protective curtain of a honey bee swarm. *Insectes Sociaux* 51, 317–324.

Czirok, A., Barabasi, A. L. & Vicsek, T. 1999. Collective motion of self-propelled particles: Kinetic phase transition in one dimension. *Physical Review Letters* 82, 209–212.

Czirok, A., Stanley, H. E. & Vicsek, T. 1997. Spontaneously ordered motion of self-propelled particles. *Journal of Physics A–Mathematical and General* 30, 1375–1385.

Czirok, A. & Vicsek, T. 2000. Collective behavior of interacting self-propelled particles. *Physica A* 281, 17–29.

Dall, S. R. X., Giraldeau, L. A., Olsson, O., McNamara, J. M. & Stephens, D. W. 2005. Information and its use by animals in evolutionary ecology. *Trends in Ecology & Evolution* 20, 187–193.

Dambach, M. & Goehlen, B. 1999. Aggregation density and longevity correlate with humidity in first-instar nymphs of the cockroach *Blattella germanica*. *Journal of Insect Physiology* 45, 423–429.

Danchin, E., Giraldeau, L. A., Valone, T. J. & Wagner, R. H. 2004. Public information: From nosy neighbors to cultural evolution. *Science* 305, 487–491.

Dawkins, R. 1976. *The selfish gene.* Oxford: Oxford University Press.

Dawkins, R. 1982. *The extended phenotype.* Oxford: W.H. Freeman and Company.

de Vries, H. & Biesmeijer, J. C. 1998. Modelling collective foraging by means of individual behaviour rules in honey-bees. *Behavioral Ecology and Sociobiology* 44, 109–124.

de Vries, H. & Biesmeijer, J. C. 2002. Self-organization in collective honeybee foraging: Emergence of symmetry breaking, cross inhibition and equal harvest-rate distribution. *Behavioral Ecology and Sociobiology* 51, 557–569.

Deneubourg, J. L., Aron, S., Goss, S. & Pasteels, J. M. 1990a. The self-organizing exploratory pattern of the Argentine ant. *Journal of Insect Behavior* 3, 159–168.

Deneubourg, J. L. & Goss, S. 1989. Collective patterns and decision-making. *Ethology, Ecology & Evolution* 1, 295–311.

Deneubourg, J. L., Goss, S., Franks, N. & Pasteels, J. M. 1989. The blind leading the blind: Modeling chemically mediated army ant raid patterns. *Journal of Insect Behavior* **2**, 719–725.

Deneubourg, J. L., Grégoire, J. C. & Le Fort, E. 1990b. Kinetics of larval gregarious behavior in the bark beetle *Dendroctonus micans* (Coleoptera: Scolytidae). *Journal of Insect Behavior* **3**, 169–182.

Deneubourg, J. L., Lioni, A. & Detrain, C. 2002. Dynamics of aggregation and emergence of cooperation. *Biological Bulletin* **202**, 262–267.

Despland, E., Rosenberg, J. & Simpson, S. J. 2004. Landscape structure and locust swarming: A satellite's eye view. *Ecography* **27**, 381–391.

Detrain, C. & Deneubourg, J. L. 2006. Self-organized structures in a superorganism: Do ants "behave" like molecules? *Physics of Life Reviews* **3**, 162–187.

Detrain, C., Tasse, O., Versaen, M. & Pasteels, J. M. 2000. A field assessment of optimal foraging in ants: Trail patterns and seed retrieval by the European harvester ant *Messor barbarus*. *Insectes Soc.* **47**, 56–62.

Devenow, A. & Welch, I. 1996. Rational herding in financial economics. *European Economic Review* **40**, 603–615.

Devigne, C. & Detrain, C. 2002. Collective exploration and area marking in the ant *Lasius niger*. *Insectes Sociaux* **49**, 357–362.

Doebeli, M. & Hauert, C. 2005. Models of cooperation based on the Prisoner's Dilemma and the Snowdrift game. *Ecology Letters* **8**, 748–766.

Doebeli, M., Hauert, C. & Killingback, T. 2004. The evolutionary origin of cooperators and defectors. *Science* **306**, 859–862.

Dornhaus, A. & Chittka, L. 2004. Why do honey bees dance? *Behavioral Ecology And Sociobiology* **55**, 395–401.

Dornhaus, A. & Franks, N. R. 2006. Colony size affects collective decision-making in the ant *Temnothorax albipennis*. *Insectes Sociaux* **53**, 420–427.

Dornhaus, A., Franks, N. R., Hawkins, R. M. & Shere, H. N. S. 2004. Ants move to improve: Colonies of *Leptothorax albipennis* emigrate whenever they find a superior nest site. *Animal Behaviour* **67**, 959–963.

Dornhaus, A., Klugl, F., Oechslein, C., Puppe, F. & Chittka, L. 2006. Benefits of recruitment in honey bees: Effects of ecology and colony size in an individual-based model. *Behavioral Ecology* **17**, 336–344.

Downs, S. G. & Ratnieks, F. L. W. 1999. Recognition of conspecifics by honeybee guards uses nonheritable cues acquired in the adult stage. *Animal Behaviour* **58**, 643–648.

Doyle, J. & Carlson, J. M. 2000. Power laws, highly optimized tolerance, and generalized source coding. *Physical Review Letters* **84**, 5656–5659.

Drent, R. & Swierstra, P. 1977. Goose flocks and food finging: Field experiments with barnacle geese in winter. *Wildfowl* **28**, 15–20.

Dugatkin, L. A. & Reeve, H. K. 1994. Behavioral ecology and levels of selection: Dissolving the group selection controversy. *Advances in the Study of Behavior* **23**, 101–133.

Dugatkin, L. A. & Reeve, H. K. 1998. *Game theory and animal behaviour.* Oxford: Oxford University Press.

Dunn, T. & Richards, M. H. 2003. When to bee social: Interactions among environmental constraints, incentives, guarding, and relatedness in a facultatively social carpenter bee. *Behavioral Ecology* **14**, 417–424.

Durier, V. & Rivault, C. 1999. Path integration in cockroach larvae, *Blattella germanica (L.)* (Insecta: Dictyoptera): Direction and distance estimation. *Animal Learning & Behavior* **27**, 108–118.

Dussutour, A., Deneubourg, J. L. & Fourcassie, V. 2005a. Amplification of individual preferences in a social context: The case of wall-following in ants. *Proceedings of the Royal Society B–Biological Sciences* **272**, 705–714.

Dussutour, A., Deneubourg, J. L. & Fourcassie, V. 2005b. Temporal organization of bi-directional traffic in the ant *Lasius niger (L.)*. *Journal Of Experimental Biology* **208**, 2903–2912.

Dussutour, A., Fourcassie, V., Helbing, D. & Deneubourg, J. L. 2004. Optimal traffic organization in ants under crowded conditions. *Nature* **428**, 70–73.

Dussutour, A., Nicolis, S. C., Despland, E. & Simpson, S. J. 2008. Individual differences influence collective behaviour in social caterpillars. *Animal Behaviour* **76**, 5–16.

Dussutour, A., Simpson, S. J., Despland, E. & Colasurdo, N. 2007. When the group denies individual nutritional wisdom. *Animal Behaviour* **74**, 931–939.

Edmonds, B., Hernandez, C. & Troitzsch, K. G. (ed.) 2008. *Social simulation technologies, advances and new discoveries*: IGI Global.

Ehrlich, P. R. & Levin, S. A. 2005. The evolution of norms. *Plos Biology* **3**, 943–948.

Emery, N. J. & Clayton, N. S. 2001. Effects of experience and social context on prospective caching strategies by scrub jays. *Nature* **414**, 443–446.

Emlen, S. T. 1997. Predicting family dynamics in social vertebrates. In *Behavioural ecology, 4th edition* (eds. J. D. Krebs, N. B. Davies). Oxford: Blackwell's Science.

Fama, E. F. 1970. Efficient capital markets—review of theory and empirical work. *Journal of Finance* **25**, 383–423.

Fama, E. F. 1991. Efficient capital-markets .2. *Journal of Finance* **46**, 1575–1617.

Farley, R. 1978. Chocolate city, vanilla suburbs. Will the trend toward racially separate communities continue? *Social Science Research* 319–44.

Farley, R. & Frey, W. H. 1994. Changes in the segregation of whites from blacks during the 1980s—small steps toward a more integrated society. *American Sociological Review* **59**, 23–45.

Farley, R., Steeh, C., Krysan, M., Jackson, T. & Reeves, K. 1994. Stereotypes and segregation—neighborhoods in the Detroit area. *American Journal of Sociology* **100**, 750–780.

Fehr, E. & Fischbacher, U. 2003. The nature of human altruism. *Nature* **425**, 785–791.

Fehr, E. & Fischbacher, U. 2004. Social norms and human cooperation. *Trends in Cognitive Sciences* **8**, 185–190.

Fernandez-Juricic, E., Erichsen, J. T. & Kacelnik, A. 2004a. Visual perception and social foraging in birds. *Trends in Ecology & Evolution* **19**, 25–31.

Fernandez-Juricic, E., Siller, S. & Kacelnik, A. 2004b. Flock density, social foraging, and scanning: An experiment with starlings. *Behavioural Ecology* **15**, 371–379.

Ferrer, J., Prats, C. & Lopez, D. 2008. Individual-based modelling: An essential tool for microbiology. *Journal of Biological Physics* **34**, 19–37.

Fewell, J. H. 1988. Energetic and time costs of foraging in harvester ants, *Pogonomyrmex occidentalis*. *Behav. Ecol. Sociobiol.* **22**, 401–408.

Fewell, J. H. 2003. Social insect networks. *Science* **301**, 1867–1870.

Fewell, J. H. & Winston, M. L. 1992. Colony state and regulation of pollen foraging in the honey bee, *Apis mellifera* L. *Behavioral Ecology and Sociobiology* **30**, 387–393.

Feynman, R. P. 1965. *The character of physical law*. London: Penguin books.

Fitzgerald, T. D. & Peterson, S. C. 1983. Elective recruitment by the eastern tent caterpillar (*Malacosoma-Americanum*). *Animal Behaviour* **31**, 417–423.

Fitzgerald, T. D. & Willer, D. E. 1983. Tent-building behavior of the eastern tent caterpillar *Malacosoma Americanum* (Lepidoptera, Lasiocampidae). *Journal of the Kansas Entomological Society* **56**, 20–31.

Fletcher, J. A. & Doebeli, M. 2006. How altruism evolves: Assortment and synergy. *Journal of Evolutionary Biology* **19**, 1389–1393.

Fletcher, J. A., Zwick, M., Doebeli, M. & Wilson, D. S. 2006. What's wrong with inclusive fitness? *Trends in Ecology & Evolution* **21**, 597–598.

Focardi, S. & Pecchioli, E. 2005 Social cohesion and foraging decrease with group size in fallow deer (*Dama dama*). *Behavioral Ecology and Sociobiology* **59**, 84–91.

Foster, K. R. 2004. Diminishing returns in social evolution: The not-so-tragic commons. *Journal of Evolutionary Biology* **17**, 1058–1072.

Foster, K. R., Wenseleers, T. & Ratnieks, F. L. W. 2006. Kin selection is the key to altruism. *Trends in Ecology & Evolution* **21**, 57–60.

Fourcassie, V. & Deneubourg, J. L. 1994. The dynamics of collective exploration and trail-formation in *Monomorium pharaonis*—experiments and model. *Physiological Entomology* **19**, 291–300.

Frank, S. A. 1998. *Foundations of social evolution*. Princeton, New Jersey: Princeton University Press.

Franks, N., Gomez, N, Goss, S and Deneubourg, J. L. 1991. The blind leading the blind in army ant raid patterns: Testing a model of self-organization (Hymenoptera : Formicidae). *Journal of Insect Behavior* **4**, 583–607.

Franks, N. R. 1989. Army ants: A collective intelligence. *Am. Sci.* **77**, 139–145.

Franks, N. R., Bryant, S., Griffiths, R. & Hemerik, L. 1990. Synchronization of the behaviour within nests of the ant *Lepothorax acervorum* (Fabricius): I. Discovering the phenomenon and its relation to the level of starvation. *Bulletin of Mathematical Biology* **52**, 597–612.

Franks, N. R. & Deneubourg, J. L. 1997. Self-organizing nest construction in ants: Individual worker behaviour and the nest's dynamics. *Animal Behaviour* **54**, 779–796.

REFERENCES

Franks, N. R., Dornhaus, A., Best, C. S. & Jones, E. L. 2006. Decision making by small and large house-hunting ant colonies: One size fits all. *Animal Behaviour* **72**, 611–616.

Franks, N. R., Dornhaus, A., Fitzsimmons, J. P. & Stevens, M. 2003a. Speed versus accuracy in collective decision making. *Proceedings of the Royal Society of London Series B–Biological Sciences* **270**, 2457–2463.

Franks, N. R., Gomez, N., Goss, S. & Deneubourg, J. L. 1991. The blind leading the blind in army ant raid patterns: Testing a model of self-organization (Hymenoptera: Formicidae). *Journal of Insect Behavior* **4**, 583–607.

Franks, N. R., Hardcastle, K. A., Collins, S., Smith, F. D., Sullivan, K. M. E., Robinson, E. J. H. & Sendova-Franks, A. B. 2008. Can ant colonies choose a far-and-away better nest over an in-the-way poor one? *Animal Behaviour* **76**, 323–334.

Franks, N. R., Hooper, J. W., Gunn, M., Bridger, T. H., Marshall, J. A. R., Gross, R. & Dornhaus, A. R. 2007. Moving targets: Collective decisions and flexible choices in house-hunting ants. *Swarm Intelligence* **1**, 81–94.

Franks, N. R., Mallon, E. B., Bray, H. E., Hamilton, M. J. & Mischler, T. C. 2003b. Strategies for choosing between alternatives with different attributes: Exemplified by house-hunting ants. *Animal Behaviour* **65**, 215–223.

Franks, N. R. & Richardson, T. 2006. Teaching in tandem-running ants. *Nature* **439**, 153.

Franks, N. R., Wilby, A., Silverman, B. W. & Tofts, C. 1992. Self-organizing nest constuction in ants: Sophisticated building by blind bulldozing. *Animal Behaviour* **44**, 357–375.

Franks, N. R. Tofts, C. & Sendova-Franks, A. 1997. Studies of the division of labour: Neither physics nor stamp collecting. *Animal Behaviour* **53**, 219–224.

Galef, B. G. & Buckley, L. L. 1996. Use of foraging trails by Norway rats. *Animal Behaviour* **51**, 765–771.

Galef, B. G. & White, D. J. 1997. Socially acquired information reduces Norway rats' latencies to find food. *Animal Behaviour* **54**, 705–714.

Galton, F. 1907. Vox Populi. *Nature* **75**, 450–451.

Ganeshaiah, K. N. & Veena, T. 1991. Topology of the foraging trails of *Leptogenys processionalis*—why are they branched? *Behav. Ecol. Sociobiol.* **29**, 263–270.

Gastner, M. T. & Newman, M. E. J. 2006. Shape and efficiency in spatial distribution networks. *Journal of Statistical Mechanics: Theory and Experiment*, 10.1088/1742-5468/2006/01/P01015.

Gautrais, J., Michelena, P., Sibbald, A., Bon, R. & Deneubourg, J. L. 2007. Allelomimetic synchronization in Merino sheep. *Animal Behaviour* **74**, 1443–1454.

George, A. J. T., Stark, J. & Chan, C. 2005. Understanding specificity and sensitivity of T-cell recognition. *Trends in Immunology* **26**, 653–659.

Gerard, J. F., Bideau, E., Maublanc, M. L., Loisel, P. & Marchal, C. 2002. Herd size in large herbivores: Encoded in the individual or emergent? *Biological Bulletin* **202**, 275–282.

Geritz, S. A. H., Kisdi, E., Meszena, G. & Metz, J. A. J. 1998. Evolutionarily singular strategies and the adaptive growth and branching of the evolutionary tree. *Evolutionary Ecology* **12**, 35–57.

Gierer, A. & Meinhard, H. 1972. Theory of biological pattern formation. *Kybernetik* **12**, 30–39.

Giraldeau, J. L. & Livoreil, B. 1996. Game theory and social foraging. In *Game theory and animal behaviour* (eds. L. A. R. Dugatkin, & H. K. Reeve): Oxford: Oxford University Press.

Giraldeau, L. A. 1988. The stable group and determinants of foraging group size. In *The ecology of social behaviour* (ed. C. N. Slobodchikoff). New York: Academic Press.

Giraldeau, L. A. 2000. *Social foraging theory*. Monographs in behavior and ecology. Princeton, New Jersey: Princeton University Press.

Giraldeau, L. A. & Beauchamp, G. 1999. Food exploitation: Searching for the optimal joining policy. *Trends in Ecology & Evolution* **14**, 102–106.

Giraldeau, L. A. & Caraco, T. 2000. *Social foraging theory*. Princeton, New Jersey: Princeton University Press.

Giraldeau, L. A. & Gillis, D. 1985. Optimal group size can be stable: A reply to Sibly. *Animal Behaviour*, 666–667.

Giraldeau, L. A. & Livoreil, B. 1998. Game theory and social foraging. In *Game theory and animal behaviour* (ed. L. A. Dugatkin & H. K. Reeve). Oxford: Oxford University Press.

Goldenfeld, N. & Kadanoff, L. P. 1999. Simple lessons from complexity. *Science* **284**, 87–89.

Gordon, D. M. 1996. The organization of work in social insect colonies. *Nature* **380**, 121–124.

Gordon, D. M. 2002. The regulation of foraging activity in red harvester ant colonies. *American Naturalist* **159**, 509–518.

Gordon, D. M. 2007. Control without hierarchy. *Nature* **446**, 143–143.

Gordon, D. M., Holmes, S. & Nacu, S. 2008. The short-term regulation of foraging in harvester ants. *Behavioral Ecology* **19**, 217–222.

Gordon, D. M. & Mehdiabadi, N. J. 1999. Encounter rate and task allocation in harvester ants. *Behavioral Ecology and Sociobiology* **45**, 370–377.

Gordon, D. M., Paul, R. E. & Thorpe, K. 1993. What is the function of encounter patterns in ant colonies. *Animal Behaviour* **45**, 1083–1100.

Gordon, D. M., Rosengren, R. & Sundstrom, L. 1992 The allocation of foragers in red wood ants. *Ecological Entomology* **17**, 114–120.

Goss, S., Aron, S., Deneubourg, J. L. & Pasteels, J. M. 1989. Self-organized shortcuts in the Argentine ant. *Naturwissenschaften* **76**, 579–581.

Goss, S. & Deneubourg, J. L. 1988. Auto-catalysis as a source of synchronized rhythmical activity in social insects. *Insectes Sociaux* **35**, 310–315.

Grafen, A. 1984. Natural selection, kin selection and group selection. In *Behavioural ecology, an evolutionary approach, 2nd edition* (eds. J. Krebs & N. Davies), pp. 62–84. Oxford: Blackwell Science.

Grafen, A. 1985. A geometric view of relatedness. In *Oxford surveys in evolutionary biology*, vol. 2 (eds. R. Dawkins & M. Ridley), pp. 28–89. Oxford: Oxford University Press.

Graham, S., Myerscough, M. R., Jones, J. C. & Oldroyd, B. P. 2006. Modelling the role of intracolonial genetic diversity on regulation of brood temperature in honey bee (*Apis mellifera* L.) colonies. *Insectes Sociaux* 53, 226–232.

Granovetter, M. 1978. Threshold models of collective behaviour. *The American Journal of Sociology* 83, 1420–1443.

Grassé, P. P. 1959. La reconstruction du nid et les coordinations interindividuelles chez *Bellicositermes natalensis* et *Cubitermes* sp. La théorie de la stigmergie: Essai d'interprétation du comportement des termites constructeurs. *Insectes Sociaux* 6, 41–83.

Greene, M. J. & Gordon, D. M. 2007a. How patrollers set foraging direction in harvester ants. *American Naturalist* 170, 943–948.

Greene, M. J. & Gordon, D. M. 2007b. Interaction rate informs harvester ant task decisions. *Behavioral Ecology* 18, 451–455.

Gregoire, G., Chate, H. & Tu, Y. H. 2003. Moving and staying together without a leader. *Physica D–Nonlinear Phenomena* 181, 157–170.

Griffin, A. S. & West, S. A. 2002. Kin selection: Fact and fiction. *Trends in Ecology & Evolution* 17, 15–21.

Griffin, A. S. & West, S. A. 2003. Kin discrimination and the benefit of helping in cooperatively breeding vertebrates. *Science* 302, 634–636.

Grimm, V. & Railsback, S. F. 2005. *Individual-based modeling and ecology*. Princeton: Princeton University Press.

Grunbaum, D. 2006. Behavior—align in the sand. *Science* 312, 1320–1322.

Grunbaum, D. & Veit, R. R. 2003. Black-browed albatrosses foraging on Antarctic krill: Density-dependence through local enhancement? *Ecology* 84, 3265–3275.

Gueron, S. 1998. The steady-state distributions of coagulation-fragmentation processes. *Journal of Mathematical Biology* 37, 1–27.

Gueron, S. & Levin, S. A. 1995. The dynamics of group formation. *Mathematical Biosciences* 128, 243–264.

Gueron, S., Levin, S. A. & Rubenstein, D. I. 1996. The dynamics of herds: From individuals to aggregations. *Journal of Theoretical Biology* 182, 85–98.

Guilford, T. & Dawkins, M. S. 1991. Receiver psychology and the evolution of animal signals. *Animal Behaviour* 42, 1–14.

Hale, J. J. 2008. Automated tracking and collective behaviour in locusts and humans. Thesis in *Zoology Department*, vol. DPhil: University of Oxford.

Hall, S. J., Wardle, C. S. & MacLennan, D. N. 1986. Predator evasion in a fish school: Test of a model for the fountain effect. *Marine biology*, 143–148.

Halley, J. D. & Burd, M. 2004. Nonequilibrium dynamics of social groups: Insights from foraging Argentine ants. *Insectes Sociaux* 51, 226–231.

Halley, J. D., Burd, M. & Wells, P. 2005. Excavation and architecture of Argentine ant nests. *Insectes Sociaux* 52, 350–356.

Hamilton, W. D. 1964. The genetical evolution of social behaviour. 1, 2. *Journal of Theoretical Biology* 7, 1–52.

Hamilton, W. D. 1971. Geometry for the selfish herd. *Journal of Theoretical Biology* **31**, 295–311.

Hardin, G. 1968. The tragedy of the commons. *Science* **162**, 1243–1248.

Hart, A. G. & Ratnieks, F. L. W. 2001. Task partitioning, division of labour and nest compartmentalisation collectively isolate hazardous waste in the leafcutting ant *Atta cephalotes*. *Behavioral Ecology and Sociobiology* **49**, 387–392.

Hart, A. G. & Ratnieks, F. L. W. 2002. Waste management in the leaf-cutting ant *Atta colombica*. *Behavioral Ecology* **13**, 224–231.

Hauert, C. 2002. Effects of space in 2 x 2 games. *International Journal of Bifurcation and Chaos* **12**, 1531–1548.

Hedström, P. 2005. *Dissecting the social*. Cambridge: Cambridge University Press.

Heinze, J., Trunzer, B., Holldobler, B. & Delabie, J. H. C. 2001. Reproductive skew and queen relatedness in an ant with primary polygyny. *Insectes Sociaux* **48**, 149–153.

Helbing, D. 2001. Traffic and related self-driven many-particle systems. *Reviews of Modern Physics* **73**, 1067–1141.

Helbing, D., Buzna, L., Johansson, A. & Werner, T. 2005. Self-organized pedestrian crowd dynamics: Experiments, simulations, and design solutions. *Transportation Science* **39**, 1–24.

Helbing, D., Farkas, I. & Vicsek, T. 2000. Simulating dynamical features of escape panic. *Nature* **407**, 487–490.

Helbing, D., Johansson, A. & Al-Abideen, H. Z. 2007. Dynamics of crowd disasters: An empirical study. *Physical Review E* **75**, 046109.

Helbing, D., Keltsch, J. & Molnár, P. 1997a. Modelling the evolution of human trail systems. *Nature* **388**, 47–50.

Helbing, D. & Molnár, P. 1995. Social force model for pedestrian dynamics. *Physical Review E* **51**, 4282–4286.

Helbing, D., Molnár, P., Farkas, I. J. & Bolay, K. 2001. Self-organizing pedestrian movement. *Environment and Planning B–Planning & Design* **28**, 361–383.

Helbing, D., Schweitzer, F., Keltsch, J. & Molnár, P. 1997b. Active walker model for the formation of human and animal trail systems. *Physical Review E* **56**, 2527–2539.

Hemelrijk, C. K. 2000. Towards the integration of social dominance and spatial structure. *Animal Behaviour* **59**, 1035–1048.

Hensor, E., Couzin, I. D., James, R. & Krause, J. 2005. Modelling density-dependent fish shoal distributions in the laboratory and field. *Oikos* **110**, 344–352.

Hoare, D. J., Couzin, I. D., Godin, J. G. J. & Krause, J. 2004. Context-dependent group size choice in fish. *Animal Behaviour* **67**, 155–164.

Hoelzer, G. A., Smith, E. & Pepper, J. W. 2006. On the logical relationship between natural selection and self-organization. *Journal of Evolutionary Biology* **19**, 1785–1794.

Hofbauer, J. & Sigmund, K. 1998. *Evolutionary games and population dynamics*. Cambridge: Cambridge University Press.

Hogendoorn, K. & Velthuis, H. H. W. 1999. Task allocation and reproductive skew in social mass provisioning carpenter bees in relation to age and size. *Insectes Sociaux* **46**, 198–207.

Holland, J. H. 1998. *Emergence from chaos to order.* Oxford: Oxford University Press.

Hölldobler, B. 1976. Recruitment behavior, home range orientation and territoriality in harvester ants, *Pogonomyrmex*. *Behav. Ecol. Sociobiol.* **1**, 3–44.

Hölldobler, B. & Möglich, M. 1980. The foraging system of *Pheidole militicida* (Hymenoptera: Formicidae). *Insectes Soc.* **27**, 237–264.

Hölldobler, B. & Wilson, E. O. 1990. *The ants.* Cambridge, Massachusetts: Belknap Press of Harvard University Press.

Hughes, B. 1971. Allelometric feeding in the domestic fowl. *British Poultry Science* **12**, 359–366.

Inada, Y. & Kawachi, K. 2002. Order and flexibility in the motion of fish schools. *Journal of Theoretical Biology* **214**, 371–387.

Irwin, A. J. & Taylor, P. D. 2000. Evolution of dispersal in a stepping-stone population with overlapping generations. *Theoretical Population Biology* **58**, 321–328.

Irwin, A. J. & Taylor, P. D. 2001. Evolution of altruism in stepping-stone populations with overlapping generations. *Theoretical Population Biology* **60**, 315–325.

Ishii, S. & Kuwahara, Y. 1968. Aggregation of German cockroach *Blattella germanica* nymphs. *Experientia* **24**, 88–89.

Jackson, D. E. & Chaline, N. 2007. Modulation of pheromone trail strength with food quality in Pharaoh's ant, *Monomorium pharaonis*. *Animal Behaviour* **74**, 463–470.

Jackson, D. E., Martin, S. J., Holcombe, M. & Ratnieks, F. L. W. 2006. Longevity and detection of persistent foraging trails in Pharaoh's ants, *Monomorium pharaonis* (L.). *Animal Behaviour* **71**, 351–359.

Janis, I. L. 1972. *Victims of groupthink.* New York: Houghton Mifflin.

Janis, I. L. 1982. *Groupthink: Psychological studies of policy decisions and fiascoes, 2nd ed.* New York: Houghton Mifflin.

Janson, S., Middendorf, M. & Beekman, M. 2005. Honeybee swarms: How do scouts guide a swarm of uninformed bees? *Animal Behaviour* **70**, 349–358.

Janson, S., Middendorf, M. & Beekman, M. 2007. Searching for a new home—couting behavior of honeybee swarms. *Behavioral Ecology* **18**, 384–392.

Jeanne, R. L. (ed.) 1988. *Interindividual behavioral variability in social insects.* Boulder, Colorado: Westview Press.

Jeanson, R., Deneubourg, J. L., Grimal, A. & Theraulaz, G. 2004a. Modulation of individual behavior and collective decision-making during aggregation site selection by the ant *Messor barbarus*. *Behavioral Ecology And Sociobiology* **55**, 388–394.

Jeanson, R., Deneubourg, J. L. & Theraulaz, G. 2004b. Discrete dragline attachment induces aggregation in spiderlings of a solitary species. *Animal Behaviour* **67**, 531–537.

Jeanson, R., Ratnieks, F. L. W. & Deneubourg, J. L. 2003. Pheromone trail decay rates on different substrates in the Pharaoh's ant, *Monomorium pharaonis*. *Physiological Entomology* **28**, 192–198.

Jeanson, R., Rivault, C., Deneubourg, J. L., Blanco, S., Fournier, R., Jost, C. & Theraulaz, G. 2005. Self-organized aggregation in cockroaches. *Animal Behaviour* **69**, 167–180.

Jerome, C. A., McInnes, D. A. & Adams, E. S. 1998. Group defense by colony-founding queens in the fire ant *Solenopsis invicta*. *Behavioral Ecology* **9**, 301–308.

Jones, J. C., Myerscough, M. R., Graham, S. & Oldroyd, B. P. 2004. Honey bee nest thermoregulation: Diversity promotes stability. *Science* **305**, 402–404.

Jones, J. C., Nanork, P. & Oldroyd, B. P. 2007. The role of genetic diversity in nest cooling in a wild honey bee, *Apis florea*. *Journal of Comparative Physiology A–Neuroethology Sensory Neural and Behavioral Physiology* **193**, 159–165.

Jost, C., Verret, J., Casellas, E., Gautrais, J., Challet, M., Lluc, J., Blanco, S., Clifton, M. J. & Theraulaz, G. 2007. The interplay between a self-organized process and an environmental template: Corpse clustering under the influence of air currents in ants. *Journal of the Royal Society Interface* **4**, 107–116.

Judd, T. M. & Sherman, P. W. 1996. Naked mole-rats recruit colony mates to food sources. *Animal Behaviour* **52**, 957–969.

Kamler, J. F., Jedrzejewska, B. & Jedrzejewski, W. 2007. Activity patterns of red deer in Bialowieza National Park, Poland. *Journal of Mammalogy* **88**, 508–514.

Kauffman, S. A. 1993. *The origins of order*. Oxford: Oxford University Press.

Keeling, M. J. & Ross, J. V. 2008. On methods for studying stochastic disease dynamics. *Journal of the Royal Society Interface* **5**, 171–181.

Keller, L. (ed.) 1999. *Levels of selection in evolution*. Princeton: Princeton University Press.

Kelley, S. 1991. The regulation of comb building in honey bee colonies. Ithaca, New York: Cornell University.

Kenne, M. & Dejean, A. 1999. Spatial distribution, size and density of nests of *Myrmicaria opaciventris* Emery (Formicidae: Myrmicinae). *Insectes Soc.* **46**, 179–185.

Kevrekidis, I. G., Gear, C. W. & Hummer, G. 2004. Equation-free: The computer-aided analysis of comptex multiscale systems. *Aiche Journal* **50**, 1346–1355.

Killingback, T. & Doebeli, M. 2002. The continuous prisoner's dilemma and the evolution of cooperation through reciprocal altruism with variable investment. *American Naturalist* **160**, 421–438.

Kimura, M. & Weiss, G. H. 1964. The stepping stone model of population structure and the decrease of the genetic correlation with distance. *Genetics* **49**, 561–576.

King, A. J. & Cowlishaw, G. 2007. When to use social information: The advantage of large group size in individual decision making. *Biology Letters* **3**, 137–139.

Kirman, A. 1993. Ants, Rationality, and Recruitment. *Quarterly Journal of Economics* **108**, 137–156.

Kitano, H. 2002. Computational systems biology. *Nature* **420**, 206–210.

Kokko, H., Johnstone, R. A. & Clutton-Brock, T. H. 2001. The evolution of cooperative breeding through group augmentation. *Proceedings of the Royal Society of London Series B–Biological Sciences* **268**, 187–196.

Kokko, H., Johnstone, R. A. & Wright, J. 2002. The evolution of parental and alloparental effort in cooperatively breeding groups: When should helpers pay to stay? *Behavioral Ecology* **13**, 291–300.

Komdeur, J. & Hatchwell, B. J. 1999. Kin recognition: Function and mechanism in avian societies. *Trends in Ecology & Evolution* **14**, 237–241.

Korb, J. 2003. Thermoregulation and ventilation of termite mounds. *Naturwissenschaften* **90**, 212–219.

Korb, J. & Heinze, J. 2004. Multilevel selection and social evolution of insect societies. *Naturwissenschaften* **91**, 291–304.

Korb, J. & Linsenmair, K. E. 1998a. The effects of temperature on the architecture and distribution of *Macrotermes bellicosus* (Isoptera, Macrotermitinae) mounds in different habitats of a West African Guinea savanna. *Insectes Sociaux* **45**, 51–65.

Korb, J. & Linsenmair, K. E. 1998b. Experimental heating of *Macrotermes bellicosus* (Isoptera, Macrotermitinae) mounds: What role does microclimate play in influencing mound architecture? *Insectes Sociaux* **45**, 335–342.

Korb, J. & Linsenmair, K. E. 2000. Thermoregulation of termite mounds: What role does ambient temperature and metabolism of the colony play? *Insectes Sociaux* **47**, 357–363.

Krause, J. 1993. The effect of Schreckstoff on the shoaling behavior of the minnow—a test of Hamilton's selfish herd theory. *Animal Behaviour* **45**, 1019–1024.

Krause, J. 1994. Differential fitness returns in relation to spatial position in groups. *Biological Reviews of the Cambridge Philosophical Society* **69**, 187–206.

Krause, J. & Ruxton, G. D. 2002. *Living in groups*. Oxford series in ecology and evolution. Oxford: Oxford University Press.

Krebs, J. R. & Davies, N. B. 1997. *Behavioural Ecology: An evolutionary approach*. Oxford, UK: Blackwell Science.

Krebs, J. R. & Davies, N. B. 1993. *An introduction to behavioural ecology*. Oxford, UK: Blackwell Science.

Kretz, T., Grunebohm, A., Kaufman, M., Mazur, F. & Schreckenberg, M. 2006a. Experimental study of pedestrian counterflow in a corridor. *Journal of Statistical Mechanics–Theory and Experiment*, 10.1088/1742-5468/2006/10/P10001.

Kretz, T., Grunebohm, A. & Schreckenberg, M. 2006b. Experimental study of pedestrian flow through a bottleneck. *Journal of Statistical Mechanics–Theory and Experiment*, 10.1088/1742-5468/2006/10/P10014.

Kretz, T., Wolki, M. & Schreckenberg, M. 2006c. Characterizing correlations of flow oscillations at bottlenecks. *Journal of Statistical Mechanics–Theory and Experiment*, 10.1088/1742-5468/2006/02/P02005.

Krugman, P. & Wells, R. 2004. *Microeconomics*. Worth publishers.

Krugman, P. W., R. 2005. *Macroeconomics*. Worth publishers.

Kuhnert, C., Helbing, D. & West, G. B. 2006. Scaling laws in urban supply networks. *Physica A–Statistical Mechanics and Its Applications* **363**, 96–103.

Kuhnholz, S. & Seeley, T. D. 1997. The control of water collection in honey bee colonies. *Behavioral Ecology and Sociobiology* **41**, 407–422.

Kuramoto, Y. 1975. Self-entrainment of a population of coupled oscillators. In *International Symposium on mathematical problems in theoretical physics*, vol. 39 (ed. H. Araki). Berlin: Springer-Verlag.

Kuramoto, Y. 1984. *Chemical Oscillations, Waves and Turbulence*. Berlin: Springer-Verlag.

Kydland, F. E. & Prescott, E. C. 1982. Time to build and aggregate fluctuations. *Econometrica* **50**, 1345–1370.

Laland, K. N. & Williams, K. 1998. Social transmission of maladaptive information in the guppy. *Behavioral Ecology* **9** 493–499.

LeBaron, B. 2006. Agent-based Computational Finance. In *Handbook of Computational Economics* (eds. L. Tesfatsion & K. Judd), pp. 1187–1232. Amsterdam: North-Holland.

Lee, S. H. 2006. Predator's attack-induced phase-like transition in prey flock. *Physics Letters A* **357**, 270–274.

Lee, S. H., Pak, H. K. & Chon, T. S. 2006. Dynamics of prey-flock escaping behavior in response to predator's attack. *Journal of Theoretical Biology* **240**, 250–259.

Lehmann, L. 2007. The evolution of trans-generational altruism: Kin selection meets niche construction. *Journal of Evolutionary Biology* **20**, 181–189.

Lehmann, L., Bargum, K. & Reuter, M. 2006. An evolutionary analysis of the relationship between spite and altruism. *Journal Of Evolutionary Biology* **19**, 1507–1516.

Lehmann, L. & Keller, L. 2006a. The evolution of cooperation and altruism—a general framework and a classification of models. *Journal of Evolutionary Biology* **19**, 1365–1376.

Lehmann, L. & Keller, L. 2006b. Synergy, partner choice and frequency dependence: Their integration into inclusive fitness theory and their interpretation in terms of direct and indirect fitness effects. *Journal of Evolutionary Biology* **19**, 1426–1436.

Leoncini, I. & Rivault, C. 2005. Could species segregation be a consequence of aggregation processes? Example of *Periplaneta americana* (L.) and *P-fuliginosa* (Serville). *Ethology* **111**, 527–540.

Levin, S. A. 1992. The problem of pattern and scale in ecology. *Ecology* **73**, 1943–1967.

Levin, S. A. 2000. *Fragile dominion: Complexity and the commons*. Cambridge, Massachusetts: Helix books.

Lieberman, E., Hauert, C. & Nowak, M. A. 2005. Evolutionary dynamics on graphs. *Nature* **433**, 312–316.

Lindauer, M. 1952. Ein Beitrag zur Frage der Arbeitsteilung im Bienenstaat. *Zeitschrift für vergleichende Physiologie* **34**, 299–345.

Lindauer, M. 1954. Temperaturregulierung und Wasserhaushalt im Bienenstaat. *Zeitschrift für vergleichende Physiologie* **36**, 391–432.

Lindauer, M. 1955. Schwarmbienen auf Wohnungssuche. *Zeitschrift für vergleichende Physiologie* **37**, 263–324.

Lindauer, M. 1961. *Communication among social bees.* New York: Atheneum.

Lioni, A. & Deneubourg, J. L. 2004. Collective decision through self-assembling. *Naturwissenschaften* **91**, 237–241.

List, C. 2004. Democracy in animal groups: A political science perspective. *Trends in Ecology & Evolution* **19**, 168–169.

Loe, L. E., Irvine, R. J., Bonenfant, C., Stien, A., Langvatn, R., Albon, S. D., Mysterud, A. & Stenseth, N. C. 2006. Testing five hypotheses of sexual segregation in an arctic ungulate. *Journal of Animal Ecology* **75**, 485–496.

Lopez, F., Acosta, F. J. & Serrano, J. M. 1994. Guerilla vs phalanx strategies of resource capture: Growth and structural plasticity in the trunk trail system of the harvester ant *Messor barbarus. J. Anim. Ecol.* **63**, 127–138.

Lucas, R. E. 1975. Equilibrium model of business cycle. *Journal of Political Economy* **83**, 1113–1144.

Lux, T. 1995. Herd behavior, bubbles and crashes. *Economic Journal* **105**, 881–896.

MacFarlane, A. M. 2006. Can the activity budget hypothesis explain sexual segregation in western grey kangaroos? *Behaviour* **143**, 1123–1143.

Mailleux, A. C., Deneubourg, J. L. & Detrain, C. 2000. How do ants assess food volume? *Animal Behaviour* **59**, 1061–1069.

Mailleux, A. C., Deneubourg, J. L. & Detrain, C. 2003a. How does colony growth influence communication in ants? *Insectes Sociaux* **50**, 24–31.

Mailleux, A. C., Deneubourg, J. L. & Detrain, C. 2003b. Regulation of ants' foraging to resource productivity. *Proceedings of the Royal Society of London Series B–Biological Sciences* **270**, 1609–1616.

Mailleux, A. C., Detrain, C. & Deneubourg, J. L. 2006. Starvation drives a threshold triggering communication. *Journal of Experimental Biology* **209**, 4224–4229.

Makse, H. A., Havlin, S. & Stanley, H. E. 1995 Modeling Urban-Growth Patterns. *Nature* **377**, 608-612.

Mallon, E. B., Pratt, S. C. & Franks, N. R. 2001. Individual and collective decision-making during nest site selection by the ant *Leptothorax albipennis. Behavioral Ecology and Sociobiology* **50**, 352–359.

Mantegna, R. N. & Stanley, H. E. 1995. Scaling behavior in the dynamics of an economic index. *Nature* **376**, 46–49.

Marzluff, J. M., Heinrich, B. & Marzluff, C. S. 1996. Raven roosts are mobile information centres. *Animal Behaviour* **51**, 89–103.

Mattila, H. R. & Seeley, T. D. 2007. Genetic diversity in honey bee colonies enhances productivity and fitness. *Science* **317**, 362–364.

REFERENCES

May, R. M. 1976. Simple mathematical models with very complicated dynamics. *Nature* **261**, 459–467.

Maynard Keynes, J. 1936. *The general theory of employment, interest and money!* New York: Harcourt.

Maynard Smith, J. 1982. *Evolution and the theory of games.* Cambridge: Cambridge University Press.

Maynard Smith, J. & Harper, D. 2005. *Animal signals.* Oxford: Oxford University Press.

Maynard Smith, J. & Szathmáry, E. 1995. *The major transitions in evolution.* Oxford: W.H. Freeman Spektrum.

McGowan, K. J. & Woolfenden, G. E. 1989. A sentinel system in the Florida scrub jay. *Animal Behaviour* **37**, 1000–1006.

Milgram, S. 1992. *The individual in the social world.* McGraw-Hill.

Milgram, S., Bickman, L. & Berkowitz, L. 1969. Note on the drawing power of crowds of different size. *Journal of Personality & Social Psychology* **13**, 79–82.

Miller, J. H. & Page, S. E. 2007. *Complex adaptive systems.* Princeton, New Jersey: Princeton University Press.

Millor, J., Amé, J. M., Halloy, J. & Deneubourg, J. L. 2006. Individual discrimination capability and collective decision-making. *Journal of Theoretical Biology* **239**, 313–323.

Milner, R. 1989. *Communication and concurrency.* Prentice Hall international series in computer science. New York: Prentice Hall.

Mirollo, R. E. & Strogatz, S. H. 1990. Synchronization of pulse-coupled biological oscillators. *Siam Journal on Applied Mathematics* **50**, 1645–1662.

Moffett, M. W. 1988. Foraging dynamics in the group-hunting myrmicine ant, *Pheidologeton diversus. Journal of Insect Behavior* **1**, 309–331.

Möglich, M. 1978. Social organization of nest emigration in *Leptothorax* (Hym., Form.). *Insectes Sociaux* **25**, 205–225.

Moody, A. L., Houston, A. I. & McNamara, J. M. 1996. Ideal free distributions under predation risk. *Behavioral Ecology and Sociobiology* **38**, 131–143.

Mottley, K. & Giraldeau, L. A. 2000. Experimental evidence that group foragers can converge on predicted producer–scrounger equilibria. *Animal Behaviour*, 341–350.

Moussaïd, M., Helbing, D., Garnier, S., Johansson, A., Combe, M. & Theraulaz, G. 2009. Experimental study of the behavioural mechanisms underlying seof-organization in human crowds. *Proceedings of the Royal Society of London Series B–Biological Sciences* **276**, 2755–2762.

Murray, J. D. 1993. *Mathematical biology.* Biomathematics. Berlin: Springer-Verlag.

Myerscough, M. R. 2003. Dancing for a decision: A matrix model for nest-site choice by honeybees. *Proceedings of the Royal Society of London Series B–Biological Sciences* **270**, 577–582.

Neda, Z., Ravasz, E., Brechet, Y., Vicsek, T. & Barabasi, A. L. 2000a. The sound of many hands clapping. *Nature* **403**, 849.

Neda, Z., Ravasz, E., Vicsek, T., Brechet, Y. & Barabasi, A. L. 2000b. Physics of the rhythmic applause. *Physical Review E* **61**, 6987–6992.

Nee, S., Colegrave, N., West, S. A. & Grafen, A. 2005. The illusion of invariant quantities in life histories. *Science* 309, 1236–1239.

Newman, M. E. J. 2003. The structure and function of complex networks. *Siam Review* 45, 167–256.

Newman, M. E. J. 2005. Power laws, Pareto distributions and Zipf's law. *Contemporary Physics* 46, 323–351.

Nicolis, G. & Prigogine, I. 1977. *Self-organization in nonequilibrium systems.* New York: John Wiley & Sons.

Nicolis, S. C. & Deneubourg, J. L. 1999. Emerging patterns and food recruitment in ants: An analytical study. *Journal of Theoretical Biology* 198, 575–592.

Nieh, J. C. 2004. Recruitment communication in stingless bees (Hymenoptera, Apidae, Meliponini). *Apidologie* 35, 159–182.

Nielsen B, L., Lawrence A, B. & Whittmore C, T. 1996. Feeding behaviour of growing pigs using single or multi-space feeders. *Applied Animal Behaviour Science* 47, 235–246.

Niwa, H. S. 1998. School size statistics of fish. *Journal of Theoretical Biology* 195, 351–361.

Niwa, H. S. 2003. Power-law versus exponential distributions of animal group sizes. *Journal of Theoretical Biology* 224, 451–457.

Niwa, H. S. 2004. Space-irrelevant scaling law for fish school sizes. *Journal of Theoretical Biology* 228, 347–357.

Nowak, M. A. & May, R. M. 1992. Evolutionary games and spatial chaos. *Nature* 359, 826–829.

Nowak, M. A. & Sigmund, K. 1998. Evolution of indirect reciprocity by image scoring. *Nature* 393, 573–577.

Nunney, L. 1985. Group selection, alturism, and structured-deme models. *American Naturalist* 126, 212–230.

Ohtsuki, H., Hauert, C., Lieberman, E. & Nowak, M. A. 2006. A simple rule for the evolution of cooperation on graphs and social networks. *Nature* 441, 502–505.

Okubo, A. 1986. Dynamical aspects of animal grouping. *Advances in Biophysics* 22, 1–94.

Okubo, A. & Chiang, H. C. 1974. An analysis of the kinematics of swarming of *Anarete pritchardi Kim* (Diptera: Cecidomyiidae). *Population Ecology*, 1–42.

Oldroyd, B. P., Gloag, R. S., Even, N., Wattanachaiyingcharoen, W. & Beekman, M. 2008. Nest site selection in the open-nesting honeybee *Apis florea*. *Behavioral Ecology and Sociobiology* 62, 1643–1653.

Ormerod, P. 1998. *Butterfly economics*. New York: Pantheon books.

Ormerod, P., Mounfield, C. & Smith, L. 2001. Non-linear modelling of burglary and violent crime in the UK. London: Volterra Consulting Ltd.

Packer, C. & Pusey, A. E. 1983. Male takeovers and female reproductive parameters—a simulation of estrous synchrony in lions (*Panthera leo*). *Animal Behaviour* 31, 334–340.

Pankiw, T. & Page, R. E. 2000. Response thresholds to sucrose predict foraging division of labor in honeybees. *Behavioral Ecology and Sociobiology* 47, 265–267.

Parrish, J. K. 1989. Reexamining the selfish herd—are central fish safer. *Animal Behaviour* **38**, 1048–1053.

Parrish, J. K., Viscido, S. V. & Grunbaum, D. 2002. Self-organized fish schools: An examination of emergent properties. *Biological Bulletin* **202**, 296–305.

Partridge, B. L. 1981. Internal dynamics and the interrelations of fish in schools. *Journal of Comparative Physiology* **144**, 313–325.

Partridge, B. L. J. 1982. The structure and function of fish schools. *Scientific American* **246**, 114–123.

Partridge, B. L. & Pitcher, T. J. 1980. The sensory basis of fish schools: Relative roles of lateral line and vision. *Journal of Comparative Physiology* **135**, 315–325.

Partridge, B. L., Pitcher, T. J., Cullen, M. J. & Wilson, J. 1980. The three-dimensional structure of fish schools. *Behavioral Ecology and Sociobiology* **6**, 277–288.

Partridge, B. L. J. 1982. The structure and function of fish schools. *Scientific American* **246**, 114–123.

Partridge, B. L. J., Johansson, J. & Kalisk, J. 1983. Structure of schools of giant bluefin tuna in Cape Cod Bay. *Environmental Biology of Fishes* **9**, 253–262.

Passino, K. M. & Seeley, T. D. 2006. Modeling and analysis of nest-site selection by honeybee swarms: The speed and accuracy trade-off. *Behavioral Ecology and Sociobiology* **59**, 427–442.

Passino, K. M., Seeley, T. D. & Visscher, P. K. 2008. Swarm cognition in honey bees. *Behavioral Ecology and Sociobiology* **62**, 401–414.

Pasteels, J. M., Deneubourg, J. L. & Goss, S. 1987. Self-organisation mechanisms in ant societies (1): Trail recruitment to newly discovered food sources. In *From individual to collective behavior in social insects* (eds. J. M. Pasteels & J. L. Deneubourg), pp. 155–175. Basel: Birkhauser Verlag.

Pedersen, J. S., Krieger, M. J. B., Vogel, V., Giraud, T. & Keller, L. 2006. Native supercolonies of unrelated individuals in the invasive Argentine ant. *Evolution* **60**, 782–791.

Pepper, J. W. & Hoelzer, G. 2001. Self-organization in biological systems. *Science* **294**, 1466–1467.

Perloff, J. M. 2007. *Microeconomics*. Boston: Pearson Education.

Perna, A., Jost, C., Couturier, E., Valverde, S., Douady, S. & Theraulaz, G. 2008a. The structure of gallery networks in the nests of termite *Cubitermes* spp. revealed by X-ray tomography. *Naturwissenschaften* **95**, 877–884.

Perna, A., Valverde, S., Gautrais, J., Jost, C., Solé, R., Kuntz, P. & Theraulaz, G. 2008b. Topological efficiency in three-dimensional gallery networks of termite nests. *Physica A: Statistical Mechanics and its Applications* **387**, 6235–6244.

Pitcher, T. J., Magurran, A. E. & Winfield, I. J. 1982. Fish in larger shoals find food faster. *Behavioral Ecology and Sociobiology* **10**, 149–151.

Portha, S., Deneubourg, J. L. & Detrain, C. 2002. Self-organized asymmetries in ant foraging: A functional response to food type and colony needs. *Behavioral Ecology* **13**, 776–781.

Portha, S., Deneubourg, J. L. & Detrain, C. 2004. How food type and brood influence foraging decisions of *Lasius niger* scouts. *Animal Behaviour* **68**, 115–122.

Potters, M., Cont, R. & Bouchaud, J. P. 1998. Financial markets as adaptive systems. *Europhysics Letters* **41**, 239–244.

Pratt, S. C. 1998a. Condition-dependent timing of comb construction by honey-bee colonies: How do workers know when to start building? *Animal Behaviour* **56**, 603–610.

Pratt, S. C. 1998b. Decentralized control of drone comb construction in honey bee colonies. *Behavioral Ecology and Sociobiology* **42**, 193–205.

Pratt, S. C. 1999. Optimal timing of comb construction by honeybee (*Apis mellifera*) colonies: A dynamic programming model and experimental tests. *Behavioral Ecology and Sociobiology* **46**, 30–42.

Pratt, S. C. 2004. Collective control of the timing and type of comb construction by honey bees (*Apis mellifera*). *Apidologie* **35**, 193–205.

Pratt, S. C. 2005a. Behavioral mechanisms of collective nest-site choice by the ant *Temnothorax curvispinosus*. *Insectes Sociaux* **52**, 383–392.

Pratt, S. C. 2005b. Quorum sensing by encounter rates in the ant *Temnothorax albipennis*. *Behavioral Ecology* **16**, 488–496.

Pratt, S. C. 2008. Efficiency and regulation of recruitment during colony emigration by the ant *Temnothorax curvispinosus*. *Behavioral Ecology and Sociobiology* **62**, 1369–1376.

Pratt, S. C., Mallon, E. B., Sumpter, D. J. T. & Franks, N. R. 2002. Quorum sensing, recruitment, and collective decision-making during colony emigration by the ant *Leptothorax albipennis*. *Behavioral Ecology and Sociobiology* **52**, 117–127.

Pratt, S. C. & Pierce, N. E. 2001. The cavity-dwelling ant *Leptothorax curvispinosus* uses nest geometry to discriminate between potential homes. *Animal Behaviour* **62**, 281–287.

Pratt, S. C. & Sumpter, D. J. T. 2006. A tunable algorithm for collective decision-making. *Proceedings of the National Academy of Sciences of the United States of America* **103**, 15906–15910.

Pratt, S. C., Sumpter, D. J. T., Mallon, E. B. & Franks, N. R. 2005. An agent-based model of collective nest choice by the ant *Temnothorax albipennis*. *Animal Behaviour* **70**, 1023–1036.

Price, D. J. D. 1976. General theory of bibliometric and other cumulative advantage processes. *Journal of the American Society for Information Science* **27**, 292–306.

Pride, E. 2005. Optimal group size and seasonal stress in ring-tailed lemurs (Lemur catta). *Behavioral Ecology* **16**, 550–560.

Prokopy, R. J. & Roitberg, B. D. 2001. Joining and avoidance behaviour in nonsocial insects. *Annual Review of Entomology*, **46**, 631–665.

Queller, D. C. 1985. Kinship, reciprocity and synergism in the evolution of social-behavior. *Nature* **318**, 366–367.

Queller, D. C. & Strassmann, J. E. 1998. Kin selection and social insects. *Bioscience* **48**, 165–175.

Queller, D. C. & Strassmann, J. E. 2006. Models of cooperation. *Journal of Evolutionary Biology* **19**, 1410–1412.

Quinn, J. L. & Cresswell, W. 2006. Testing domains of danger in the selfish herd: Sparrowhawks target widely spaced redshanks in flocks. *Proceedings of the Royal Society B–Biological Sciences* **273**, 2521–2526.

Radakov, D. V. 1973. *Schooling in the ecology of fish.* New York: John Wiley & Sons.

Raffa, K. F. 2001. Mixed messages across multiple trophic levels: The ecology of bark beetle chemical communication systems. *Chemoecology* **11**, 49–65.

Rands, S. A., Cowlishaw, G., Pettifor, R. A., Rowcliffe, J. M. & Johnstone, R. A. 2003. Spontaneous emergence of leaders and followers in foraging pairs. *Nature* **423**, 432–434.

Rangel, J. & Seeley, T. D. 2008. The signals initiating the mass exodus of a honeybee swarm from its nest. *Animal Behaviour* **76**, 1943–1952.

Ranta, E., Peuhkuri, N., Laurila, A., Rita, H. & Metcalfe, N. B. 1996. Producers, scroungers and foraging group structure. *Animal Behaviour* **51**, 171–175.

Ratcliffe, J. M. & Hofstedeter, H. M. 2005. Roosts as information centres: Social learning of food preferences in bats. *Biology Letters* **1**, 72–74.

Ratnieks, F. L. W. 2006. The evolution of cooperation and altruism: The basic conditions are simple and well known. *Journal of Evolutionary Biology* **19**, 1413–1414.

Ratnieks, F. L. W., Foster, K. R. & Wenseleers, T. 2006. Conflict resolution in insect societies. *Annual Review of Entomology* **51**, 581–608.

Rayor, L. S. & Uetz, G. W. 1990. Trade-offs in foraging success and predation risk with spatial position in colonial spiders. *Behavioral Ecology and Sociobiology* **27**, 77–85.

Reed, W. J. 2001. The Pareto, Zipf and other power laws. *Economics Letters* **74**, 15–19.

Reeve, H. K. & Keller, L. 2001. Tests of reproductive-skew models in social insects. *Annual Review of Entomology* **46**, 347–385.

Reeve, H. K., Westneat, D. F., Noon, W. A., Sherman, P. W. & Aquadro, C. F. 1990. DNA fingerprinting reveals high-levels of inbreeding in colonies of the eusocial naked mole-rat. *Proceedings of the National Academy of Sciences of the United States of America* **87**, 2496–2500.

Reynolds, C. W. 1987. Flocks, herds and schools: A distributed behavioural model. *Computer Graphics* **21**, 25–33.

Richmond, P. 2001. Power law distributions and dynamic behaviour of stock markets. *European Physical Journal B* **20**, 523–526.

Richner, H. & Heeb, P. 1996. Communal life: Honest signaling and the recruitment center hypothesis. *Behavioral Ecology* **7**, 115–118.

Riley, J. R., Greggers, U., Smith, A. D., Reynolds, D. R. & Menzel, R. 2005. The flight paths of honeybees recruited by the waggle dance. *Nature* **435**, 205–207.

Rittschof, C. C. & Seeley, T. D. 2008. The buzz-run: How honeybees signal "Time to go!" *Animal Behaviour* **75**, 189–197.

Rivault, C. 1989. Spatial distribution of the cockroach, *Blattella germanica*, in a swimming-bath facility. *Entomologia Experimentalis et Applicata* 53, 247–255.

Rivault, C. & Cloarec, A. 1998. Cockroach aggregation: Discrimination between strain odours in *Blattella germanica*. *Animal Behaviour* 55, 177–184.

Rivault, C., Cloarec, A. & Sreng, L. 1998. Cuticular extracts inducing aggregation in the German cockroach, *Blattella germanica* (L.). *Journal of Insect Physiology* 44, 909–918.

Robinson, E. J. H., Jackson, D. E., Holcombe, M. & Ratnieks, F. L. W. 2005. Insect communication—"no entry" signal in ant foraging. *Nature* 438, 442–442.

Rook A, J. & Penning P, D. 1991. Synchronization of eating, ruminating and idling activity by grazing sheep. *Applied Animal Behaviour Science* 32, 157–166.

Rosengren, R. & Sundström, L. 1987. The foraging system of a red wood ant colony (*Formica s. str.*)—collecting and defending food through an extended phenotype. In *From individual to collective behavior in social insects: les Treilles Workshop. Experientia Supplementum*, vol. 54 (eds. J. M. Pasteels & J. L. Deneubourg), pp. 117–137. Basel: Birkhauser.

Rousset, F. 2004. *Genetic structure and selection in subdivided populations.* Princeton, New Jersey: Princeton University Press.

Ruckstuhl, K. E. 1999. To synchronise or not to synchronise: A dilemma for young bighorn males? *Behaviour* 136, 805–818.

Ruckstuhl, K. E. & Neuhaus, P. 2001. Behavioral synchrony in ibex groups: Effects of age, sex and habitat. *Behaviour* 138, 1033–1046.

Ruckstuhl, K. E. & Neuhaus, P. 2002. Sexual segregation in ungulates: A comparative test of three hypotheses. *Biological Reviews* 77, 77–96.

Ruxton, G. D., Fraser, C. & Broom, M. 2005. An evolutionarily stable joining policy for group foragers. *Behavioral Ecology* 16, 856–864.

Saam, N. J. & Sumpter, D. 2008. EU institutional reforms—how do member states reach a decision? *Journal of Policy Modeling* 30, 71–86.

Saffre, F., Mailleux, A. C. & Deneubourg, J. L. 2000. Exploratory recruitment plasticity in a social spider (*Anelosimus eximius*). *Journal Of Theoretical Biology* 205, 37–46.

Santos, F. C. & Pacheco, J. M. 2006. A new route to the evolution of cooperation. *Journal of Evolutionary Biology* 19, 726–733.

Savit, R., Manuca, R. & Riolo, R. 1999. Adaptive competition, market efficiency, and phase transitions. *Physical Review Letters* 82, 2203–2206.

Schelling, T. C. 1969. Models of segregation. *American Economic Review* 59, 488–493.

Schelling, T. C. 1971. Dynamic models of segregation. *Journal of Mathematical Sociology* 1, 143–186.

Schelling, T. C. 1978. *Micromotives and macrobehavior.* New York: W. W. Norton and Company.

Schmickl, T. & Crailsheim, K. 2002. How honeybees (*Apis mellifera* L.) change their broodcare behavior in response to non-foraging conditions and poor pollen conditions. *Behavioral Ecology Sociobiology* 51, 415–425.

Schmickl, T. & Crailsheim, K. 2008a. An individual-based model of task selection in honeybees. *From Animals to Animats 10, Proceedings* **5040**, 383–392.

Schmickl, T. & Crailsheim, K. 2008b. TaskSelSim: A model of the self-organization of the division of labour in honeybees. *Mathematical and Computer Modelling of Dynamical Systems* **14**, 101–125.

Schneirla, T. C. 1971. *Army ants. A study in social organization*. San Francisco: W. H. Freeman & Co.

Schuessler, R. 1989. Exit threats and cooperation under anonymity. *Journal of Conflict Resolution* **33**, 728–749.

Schultz, K. M., Passino, K. M. & Seeley, T. D. 2008. The mechanism of flight guidance in honeybee swarms: Subtle guides or streaker bees? *Journal of Experimental Biology* **211**, 3287–3295.

Seeley, T. D. 1977. Measurement of nest cavity volume by the honey bee (*Apis mellifera*). *Behavioral Ecology and Sociobiology* **2**, 201–227.

Seeley, T. D. 1983. Division of labour between scouts and recruits in honeybee foraging. *Behavioral Ecology and Sociobiology* **12**, 253–259.

Seeley, T. D. 1992. The tremble dance of the honey bee: Message and meanings. *Behavioral Ecology and Sociobiology* **31**, 375–383.

Seeley, T. D. 1995. *The wisdom of the hive*. Cambridge, Massachusetts: Belknap Press of Harvard University Press.

Seeley, T. D. 1997. Honey bee colonies are group-level adaptive units. *American Naturalist* **150**, S22–S41.

Seeley, T. D. 1998. Thoughts on information and integration in honey bee colonies. *Apidologie* **29**, 67–80.

Seeley, T. D. 2002. When is self-organization used in biological systems? *Biological Bulletin* **202**, 314–318.

Seeley, T. D. 2003. Consensus building during nest-site selection in honey bee swarms: The expiration of dissent. *Behavioral Ecology and Sociobiology* **53**, 417–424.

Seeley, T. D. & Buhrman, S. C. 1999. Group decision making in swarms of honey bees. *Behavioral Ecology and Sociobiology* **45**, 19–31.

Seeley, T. D. & Buhrman, S. C. 2001. Nest-site selection in honey bees: How well do swarms implement the "best-of-N" decision rule? *Behavioral Ecology and Sociobiology* **49**, 416–427.

Seeley, T. D., Camazine, S. & Sneyd, J. 1991. Collective decision-making in honey bees: How colonies choose among nectar sources. *Behavioral Ecology and Sociobiology* **28**, 277–290.

Seeley, T. D., Kleinhenz, M., Bujok, B. & Tautz, J. 2003. Thorough warm-up before take-off in honey bee swarms. *Naturwissenschaften* **90**, 256–260.

Seeley, T. D., Morse, R. A. & Visscher, P. K. 1979. The natural history of the flight of honey bee swarms. *Psyche* **86**, 103–113.

Seeley, T. D. & Tautz, J. 2001. Worker piping in honey bee swarms and its role in preparing for liftoff. *Journal of Comparative Physiology A–Sensory Neural and Behavioral Physiology* **187**, 667–676.

Seeley, T. D. & Towne, W. F. 1992. Tactics of dance choice in honey bees: Do foragers compare dances? *Behavioral Ecology and Sociobiology* **30**, 59–69.

Seeley, T. D. & Visscher, P. K. 2003. Choosing a home: How the scouts in a honey bee swarm perceive the completion of their group decision making. *Behavioral Ecology and Sociobiology* 54, 511–520.

Seeley, T. D. & Visscher, P. K. 2004a. Group decision making in nest-site selection by honey bees. *Apidologie* 35, 101–116.

Seeley, T. D. & Visscher, P. K. 2004b. Quorum sensing during nest-site selection by honeybee swarms. *Behavioral Ecology and Sociobiology* 56, 594–601.

Seeley, T. D. & Visscher, P. K. 2008. Sensory coding of nest-site value in honeybee swarms. *Journal of Experimental Biology* 211, 3691–3697.

Seeley, T. D., Weidenmuller, A. & Kuhnholz, S. 1998. The shaking signal of the honey bee informs workers to prepare for greater activity. *Ethology* 104, 10–26.

Sethna, J. P. 2006. *Entropy, order parameters and complexity*: Oxford University Press.

Sharp, S. P., McGowan, A., Wood, M. J. & Hatchwell, B. J. 2005. Learned kin recognition cues in a social bird. *Nature* 434, 1127–1130.

Shepherd, J. D. 1982. Trunk trails and the searching strategy of a leaf-cutter ant, *Atta colombica*. *Behav. Ecol. Sociobiol.* 11, 77–84.

Sherman, P. W. V., P. K. 2002. Honeybee colonies gain fitness through dancing. *Nature* 419, 920–922.

Shiller, R. J. 2000. *Irrational exuberance*. Princeton, New Jersey: Princeton University Press.

Shiller, R. J. 2003. From efficient markets theory to behavioral finance. *Journal of Economic Perspectives* 17, 83–104.

Sibly, R. M. 1983. Optimal group size is unstable. *Animal Behaviour* 31, 947–948.

Simon, H. A. 1955. On a class of skew distribution functions. *Biometrika* 42, 425–440.

Simons, A. M. 2004. Many wrongs: The advantage of group navigation. *Trends in Ecology & Evolution* 19, 453–455.

Sinclair, A. R. E. 1977. *The American buffalo*. Chicago: University of Chicago Press.

Sjoberg, M., Albrectsen, B. & Hjalten, J. 2000. Truncated power laws: A tool for understanding aggregation patterns in animals? *Ecology Letters* 3, 90–94.

Skog, O. J. 1986. The long waves of alchohol consumption: A social network perspective of cultural change. *Social networks* 8, 1–32.

Skyrms, B. 2004. *The stag hunt and the evolution of social structure*. Cambridge: Cambridge University Press.

Sole, R. V. & Miramontes, O. 1995. Information at the edge of chaos in fluid neural networks. *Physica D* 80, 171–180.

Sole, R. V., Miramontes, O. & Goodwin, B. C. 1993. Oscillations and chaos in ant societies. *Journal of Theoretical Biology* 161, 343–357.

Sonerud, G. A., Smedshaug, C. A. & Brathen, O. 2001. Ignorant hooded crows follow knowledgeable roost-mates to food: Support for the information centre hypothesis. *Proceedings of the Royal Society of London Series B–Biological Sciences* 268, 827–831.

Sornette, D. 2003a. Critical market crashes. *Physics Reports–Review Section of Physics Letters* 378, 1–98.

Sornette, D. 2003b. *Why stock markets crash*. Princeton, New Jersey: Princeton University Press.

Sornette, D. 2004. *Critical phenomena in natural sciences*. Springer Verlag.

Stainforth, D. A., Aina, T., Christensen, C., Collins, M., Faull, N., Frame, D. J., Kettleborough, J. A., Knight, S., Martin, A., Murphy, J. M., Piani, C., Sexton, D., Smith, L. A., Spicer, R. A., Thorpe, A. J. & Allen, M. R. 2005. Uncertainty in predictions of the climate response to rising levels of greenhouse gases. *Nature* 433, 403–406.

Stanley, H. E., Amaral, L. A. N., Gabaix, X., Gopikrishnan, P. & Plerou, V. 2001. Quantifying economic fluctuations. *Physica A–Statistical Mechanics and Its Applications* 302, 126–137.

Stern, K. & McClintock, M. K. 1998. Regulation of oculation by human pheromones. *Nature* 392, 177–179.

Strogatz, S. 2003. Sync: The emerging science of spontaneous order. New York: Hyperion.

Strogatz, S. H. 1994. *Nonlinear Dynamics and Chaos*. Reading, Massachusetts: Addison-Wesley.

Strogatz, S. H. 2000. From Kuramoto to Crawford: Exploring the onset of synchronization in populations of coupled oscillators. *Physica D* 143, 1–20.

Strogatz, S. H. 2001. Exploring complex networks. *Nature* 410, 268–276.

Strogatz, S. H. & Stewart, I. 1993. Coupled oscillators and biological synchronization. *Scientific American* 269, 102–109.

Sumpter, D. J. T. 2006. The principles of collective animal behaviour. *Philosophical Transactions of the Royal Society of London Series B–Biological Sciences* 361, 5–22.

Sumpter, D. J. T. & Beekman, M. 2003. From nonlinearity to optimality: Pheromone trail foraging by ants. *Animal Behaviour* 66, 273–280.

Sumpter, D. J. T., Blanchard, G. B. & Broomhead, D. S. 2001. Ants and agents: A process algebra approach to modelling ant colony behaviour. *Bulletin of Mathematical Biology* 63, 951–980.

Sumpter, D. J. T. & Brännström, Å. 2008. Synergy in social communication. In *Sociobiology of Communication* (eds. P. D'Ettorre & D. Hughes). Oxford: Oxford Univeristy Press.

Sumpter, D. J. T. & Broomhead, D. S. 2000. Shape and dynamics of thermoregulating honey bee clusters. *Journal of Theoretical Biology* 204, 1–14.

Sumpter, D. J. T. & Broomhead, D. S. 2001. Relating individual behaviour to population dynamics. *Proceedings of the Royal Society of London Series B–Biological Sciences* 268, 925–932.

Sumpter, D. J. T., Buhl, J., Biro, D. & Couzin, I. D. 2008a. Information transfer in moving animal groups. *Theory in Biosciences* 127, 177–186.

Sumpter, D. J. T., Krause, J., James, R., Couzin, I. D. & Ward, A. J. W. 2008b. Consesus decision-making by fish. *Current Biology* 18, 1773–1777.

Sumpter, D. J. T. & Pratt, S. C. 2003. A modelling framework for understanding social insect foraging. *Behavioral Ecology and Sociobiology* 53, 131–144.

Sumpter, D. J. T. & Pratt, S. C. 2008. Quorum responses and consensus decision-making. *Philosophical Transactions of the Royal Society of London: Series B*, in press.

Surowiecki, J. 2004. *The wisdom of crowds*. London: Little, Brown.

Szabo, B., Szollosi, G. J., Gonci, B., Juranyi, Z., Selmeczi, D. & Vicsek, T. 2006. Phase transition in the collective migration of tissue cells: Experiment and model. *Physical Review E* **74**, 061908.

Takayasu, H. 1989. Steady-state distribution of generalized aggregation system with injection. *Physical Review Letters* **63**, 2563–2565.

Takayasu, H., Nishikawa, I. & Tasaki, H. 1988. Power-law mass-distribution of aggregation systems with injection. *Physical Review A* **37**, 3110–3117.

Takayasu, H., Takayasu, M., Provata, A. & Huber, G. 1991. Statistical properties of aggregation with injection. *Journal of Statistical Physics* **65**, 725–745.

Tamm, S. 1980. Bird orientation—single homing pigeons compared with small flocks. *Behavioral Ecology and Sociobiology* **7**, 319–322.

Taylor, P. D. 1992a. Altruism in viscous populations—n inclusive fitness model. *Evolutionary Ecology* **6**, 352–356.

Taylor, P. D. 1992b. Inclusive fitness in a homogeneous environment. *Proceedings of the Royal Society of London Series B–Biological Sciences* **249**, 299–302.

Taylor, P. D. & Frank, S. A. 1996. How to make a kin selection model. *Journal of Theoretical Biology* **180**, 27–37.

Taylor, P. D. & Irwin, A. J. 2000. Overlapping generations can promote altruistic behavior. *Evolution* **54**, 1135–1141.

Templeton, J. J. & Giraldeau, L. A. 1996. Vicarious sampling: The use of personal and public information by starlings foraging in a simple patch environment. *Behavioral Ecology and Sociobiology* **38**, 105–114.

Templeton, J. J. & Giraldeau, L. A. 1995. Public information cues affect the scrounging decisions of starlings. *Animal Behaviour* **49**, 1617–1626.

Theraulaz, G., Bonabeau, E., Nicolis, S. C., Sole, R. V., Fourcassie, V., Blanco, S., Fournier, R., Joly, J. L., Fernandez, P., Grimal, A., Dalle, P. & Deneubourg, J. L. 2002. Spatial patterns in ant colonies. *Proceedings of the National Academy of Sciences of the United States of America* **99**, 9645–9649.

Theraulaz, G., Gautrais, J., Camazine, S. & Deneubourg, J. L. 2003. The formation of spatial patterns in social insects: From simple behaviours to complex structures. *Philosophical Transactions of the Royal Society of London Series A–Mathematical Physical and Engineering Sciences* **361**, 1263–1282.

Thom, C., Gilley, D. C., Hooper, J. & Esch, H. E. 2007. The scent of the waggle dance. *Plos Biology* **5**, 1862–1867.

Tibbetts, E. A. & Reeve, H. K. 2003. Benefits of foundress associations in the paper wasp *Polistes dominulus*: Increased productivity and survival, but no assurance of fitness returns. *Behavioral Ecology* **14**, 510–514.

Tien, J. H., Levin, S. A. & Rubenstein, D. I. 2004. Dynamics of fish shoals: Identifying key decision rules. *Evolutionary Ecology Research* **6**, 555–565.

Tofts, C. 1991. Describing social insect behaviour using process algebra. *Transactions of the Society for Computer Simulation* **December**, 227–283.

Tofts, C. 1993. Algorithms for task allocation in ants. (A study of temporal polyethism: Theory). *Bulletin of Mathematical Biology* 55, 891–918.

Tofts, C. M. 1994. Processes with probabilities, priorities and time. *Formal Aspects of Computer Science* 6, 536–564.

Topoff, H. 1984. Social organization of raiding and emigrations in army ants. *Advances in the Study of Behavior* 14, 81–126.

Treherne, J. E. & Foster, W. A. 1981. Group transmission of predator avoidance behaviour in a marine insect: The Trafalgar effect. *Animal Behaviour* 28, 911–917.

Trivers, R. L. 1971. Evolution of reciprocal altruism. *Quarterly Review of Biology* 46, 35–57.

Tschinkel, W. R. 2004. The nest architecture of the Florida harvester ant, *Pogonomyrmex badius*. *Journal of Insect Science* 4, 1–19.

Turing, A. M. 1952. The Chemical basis of morphogenesis. *Philosophical Transactions of the Royal Society of London Series B–Biological Sciences* 237, 37–72.

Uvarov, B. P. 1977. Grasshoppers and locusts. Vol. 2. Cambridge: Cambridge University Press.

van Baalen, M. & Rand, D. A. 1998. The unit of selection in viscous populations and the evolution of altruism. *Journal Of Theoretical Biology* 193, 631–648.

Vasconcellos, H. L. 1990. Foraging activity of two species of leaf-cutting ants (*Atta*) in a primary forest of the central Amazon. *Insectes Soc.* 37, 131–145.

Vercelli, A. 1991. *Methodological foundations of macroeconomics: Keynes and Lucas*. Cambridge: Cambridge University Press.

Vickery, W. L., Giraldeau, L. A., Templeton, J. J., Kramer, D. L. & Chapman, C. A. 1991. Producers, scroungers, and group foraging. *American Naturalist* 137, 847–863.

Vicsek, T., Czirok, A., Benjacob, E., Cohen, I. & Shochet, O. 1995. Novel type of phase-transition in a system of self-driven particles. *Physical Review Letters* 75, 1226–1229.

Vincent, T. L. & Brown, J. S. 1984. Stability in an evolutionary game. *Theoretical Population Biology*, 408–427.

Vine, I. 1971. Risk of visual detection and pursuit by a predator and the selective advantage of flocking behaviour. *Journal of Theoretical Biology* 30, 405–422.

Visscher, P. K. 2007. Nest-site selection and group decision-making in social insects. *Annual Review of Entomology* 52, in press.

Visscher, P. K. & Camazine, S. 1999. Collective decisions and cognition in bees. *Nature* 397, 400.

Visscher, P. K. & Seeley, T. D. 2007. Coordinating a group departure: Who produces the piping signals on honeybee swarms? *Behavioral Ecology and Sociobiology* 61, 1615–1621.

Vittori, K., Talbot, G., Gautrais, J., Fourcassie, V., Araujo, A. F. R. & Theraulaz, G. 2006. Path efficiency of ant foraging trails in an artificial network. *Journal of Theoretical Biology* 239, 507–515.

von Bertalanffy, L. 1968. *General system theory.* New York: George Braziller.

von Dassow, G., Meir, E., Munro, E. & Odell, G. 2000. The segment polarity network is a robust development module. *Nature* **406**, 188–192.

von Frisch, K. 1967. *The Dance language and orientation of bees.* Cambridge, Massachusetts: The Belknap Press of Harvard University Press.

Wagner, R. H. & Danchin, E. 2003. Conspecific copying: A general mechanism of social aggregation. *Animal Behaviour* **65**, 405–408.

Wahl, L. M. & Nowak, M. A. 1999. The continuous prisoner's dilemma: 1. Linear reactive strategies. *Journal of Theoretical Biology* **200**, 307–321.

Wallin, K. F. & Raffa, K. F. 2004. Feedback between individual host selection behavior and population dynamics in an eruptive herbivore. *Ecological Monographs* **74**, 101–116.

Wallraff, H. G. 1978. Social interrelations involved in migratory orientation of birds—possible contribution of field studies. *Oikos* **30**, 401–404.

Walter, A. & Weber, F. M. 2006. Herding in the German mutual fund industry. *European Financial Management* **12**, 375–406.

Ward, A., Sumpter, D. J. T., Couzin, I. D., Hart & Krause, J. 2008. Quorum decision-making facilitates information transfer in fish schools. *Proceedings of the National Academy of Sciences of the United States of America*, 6948–6953.

Watts, D. J. & Strogatz, S. H. 1998. Collective dynamics of "small-world" networks. *Nature* **393**, 440–442.

Weber, N. A. 1972. The fungus-culturing behavior of ants. *Am. Zool.* **12**, 577–587.

Weeks, C. A. D., Davies, H. C. , Hunt, P. & Kestin, S. C. 2000. The behaviour of broiler chickens and its modification by lameness. *Applied Animal Behaviour Science* **67**, 111–125.

Weidenmuller, A. 2004. The control of nest climate in bumblebee (*Bombus terrestris*) colonies: Interindividual variability and self reinforcement in fanning response. *Behavioural Ecology* **15**, 120–128.

Weimerskirch, H., Martin, J., Clerquin, Y., Alexandre, P. & Jiraskova, S. 2001. Energy saving in flight formation—pelicans flying in a "V" can glide for extended periods using the other birds' air streams. *Nature* **413**, 697–698.

Welch, I. 2000. Herding among security analysts. *Journal of Financial Economics* **58**, 369–396.

Wertheim, B., van Baalen, E. J. A., Dicke, M. & Vet, L. E. M. 2005. Pheromone-mediated aggregation in nonsocial arthropods: An evolutionary ecological perspective. *Annual Review of Entomology* **50**, 321–346.

West, S. A., Gardner, A. & Griffin, A. S. 2006. Altruism. *Current Biology* **16**, R482–R483.

West, S. A., Griffin, A. S. & Gardner, A. 2007. Social semantics: Altruism, cooperation, mutualism, strong reciprocity and group selection. *Journal of Evolutionary Biology* **20**, 415–432.

West, S. A., Murray, M. G., Machado, C. A., Griffin, A. S. & Herre, E. A. 2001. Testing Hamilton's rule with competition between relatives. *Nature* **409**, 510–513.

White, D. J. & Galef, B. G. 1999a. Affiliative preferences are stable and predict mate choices in both sexes of Japanese quail, *Coturnix japonica*. *Animal Behaviour* **58**, 865–871.

White, D. J. & Galef, B. G. 1999b. Social effects on mate choices of male Japanese quail, *Coturnix japonica*. *Animal Behaviour* **57**, 1005–1012.

White, D. J. & Galef, B. G. 2000. "Culture" in quail: Social influences on mate choices of female *Coturnix japonica*. *Animal Behaviour* **59**, 975–979.

Wiener, N. 1948. *Cybernetics*. New York: John Wiley & Sons.

Wilson, D. S. 1983. The group selection controversy: History and current status. *Annual Review of Ecology and Systematics* **14**, 159–187.

Wilson, D. S. & Dugatkin, L. A. 1997. Group selection and assortative interactions. *American Naturalist* **149**, 336–351.

Wilson, E. O. 1971. *The Insect Societies*. Cambridge, Massachusetts: Belknap Press of Harvard University Press.

Wilson, E. O. & Holldobler, B. 2005. Eusociality: Origin and consequences. *Proceedings of the National Academy of Sciences of the United States of America* **102**, 13367–13371.

Winston, M. L. 1987. *The Biology of the honey bee*. Cambridge, Massachusetts: Harvard University Press.

Wolfram, S. 2002. *A new kind of science*. Wolfram Media.

Wood, A. J. & Ackland, G. J. 2007. Evolving the selfish herd: Emergence of distinct aggregating strategies in an individual-based model. *Proceedings of the Royal Society B–Biological Sciences* **274**, 1637–1642.

Wright, J., Stone, R. E. & Brown, N. 2003. Communal roosts as structured information centres in the raven, *Corvus corax*. *Journal of Animal Ecology* **72**, 1003–1014.

Wright, S. 1943. Isolation by distance. *Genetics* **28**, 114–138.

Wyatt, T. D. 2003. *Pheromones and animal behaviour*. Cambridge: Cambridge University Press.

Zahavi, A. 1971. The function of pre-roost gatherings and communal roosts. *Ibis* **113**, 106–109.

Zeeman, E. C. 1981. Dynamics of the evolution of animal conflicts. *Journal of Theoretical Biology* **89**, 249–270.

Zheng, M., Kashimori, Y., Hoshino, O., Fujita, K. & Kambara, T. 2005. Behavior pattern (innate action) of individuals in fish schools generating efficient collective evasion from predation. *Journal of Theoretical Biology* **235**, 153–167.

Index

Note: Page numbers in **bold** indicate boxes; *italic* page numbers refer to illustrations including charts and graphs.